Plants and Beekeeping

Plants
and Beekeeping

*an account of those plants,
wild and cultivated,
of value to the hive bee, and for
honey production in the
British Isles*

F. N. HOWES
D.Sc.

with a Foreword by Eva Crane, M.Sc., Ph.D.,
Director, International Bee Research Association

FABER AND FABER
London & Boston

First published in 1945
by Faber and Faber Limited
This new and revised edition published in 1979
by Faber and Faber Limited
3 Queen Square, London W.C. 1
Printed in Great Britain by
Latimer Trend & Company Ltd Plymouth
All rights reserved

British Library Cataloguing in Publication Data

Howes, Frank Norman
    Plants and beekeeping. – New and revised ed.
    1. Honey plants – Great Britain
    I. Title
    581   SF535.2.G7

    ISBN 0–571–04987–7
    ISBN 0–571–11358–3 Pbk

Contents

Section 3
OTHER PLANTS VISITED BY THE HONEY BEE FOR NECTAR OR POLLEN
page 96

Notes on the 1979 Edition

Bracketed figures in the text refer to items of the Bibliography, page 219, which appeared in the first edition, and also (items 31 onwards) of the Supplementary Bibliography, page 221, which makes accessible information published since 1945.

Minor amendments only have been made to the text of the 1945 edition; these are mainly concerned with the updating of plant names. Former or alternative names are shown in brackets.

The majority of plant names shown in the text in roman type are those listed under individual headings in Sections 2 and 3, and they will be found in the General Index. The plant names shown in the text in italic type are those which appear in the Index of Genera.

Preface

There has been a marked increase of interest in beekeeping and the production of honey throughout the country in recent years. This may have been initiated by the Second World War, with the consequent shortage of sweetening materials, and partly by other considerations, such as the better understanding of some of the major bee diseases that now prevails. The number of beekeepers has been doubled or trebled in many localities according to the statistics of Beekeepers' Associations and doubtless the total production of home-produced honey has been stepped up considerably. It is to be hoped this increase in the Nation's annual honey crop will continue, and, what is of even greater importance, that this increase in the nation's bee population will also be maintained, for it has been proved that the main value of the honey bee in the national economy is as a pollinator for fruit, clovers, and other seed and farm crops. Its value in this respect far outweighs its value as a producer of honey.

Plant nectar has been described as the raw material of the honey industry and those plants that produce it, in a manner available to the honey bee, constitute the very foundations of apiculture. They are obviously of first importance to the beekeeper, whether he or she is a large- or small-scale beekeeper or belongs to the hobbyist class. A knowledge of these plants and their relative values, for nectar or for pollen, is likely to add much to the pleasure and the profit of beekeeping An attempt has here been made to deal with the more important bee plants in the British Isles as well as many others that are only of minor importance. Among the latter are to be found both wild and garden plants. Although not sufficiently prevalent in most cases to affect honey yields to any extent such plants have been purposely included in the knowledge that their presence is always beneficial, especially as they so often help to maintain or support bees between the major nectar flows. Much of the pollen collected by bees, so vital for the sustenance of their young, comes from such plants. Furthermore, beekeepers are often keen gardeners and nature lovers and interested in any plant that proves attractive to bees. This no doubt accounts for the present popularity of bee gardens or gardens devoted exclusively to the cultivation of good bee plants, to which a chapter has been given. From the earliest times, gardening has been closely associated or connected with beekeeping and the two are obviously complementary and and well suited for being carried on together.

Many owners of gardens and flower lovers with no special interest in beekeeping derive great pleasure from observing bees industriously at work on flowers and are fond of growing some of those plants which they know will prove a special attraction, even though may not always be in the front rank as garden plants. Indications are given as to what plants are likely to be most suitable in this connection and special emphasis laid on some of the newer plant introductions.

Among the minor bee plants will be found quite a number of introduced trees and shrubs that are grown to a greater or lesser extent for ornament. Some of these are important for honey in their native land and where this is known the fact is mentioned. As some of these plants, especially among those from the Orient, are of comparatively recent introduction, they may become more generally grown and therefore more useful as bee fodder at some future time. It is for this reason they have been included.

The more serious-minded beekeeper and honey producer may be interested only in those plants that fill or help to fill his hives. These will be found described at much greater length in Section 2. Some of the major honey plants of Britain, such as the clovers, lime, heather and fruit trees are also important for honey in other countries. It is hoped, therefore, that the book may not be without interest to beekeepers and those interested in such plants in other lands.

The writer is indebted to colleagues and fellow beekeepers for helpful suggestions in the preparation of this work, which has been in the course of preparation for many years. During this period much time has been spent in observing the behaviour of the honey bee towards various wild and introduced plants at different seasons of the year and in different parts of the country.

F. N. HOWES

14 Nylands Avenue
Kew, Surrey
June 1945

Foreword to the 1979 Edition

by Dr Eva Crane
Director, International Bee Research Association

Many people will be pleased to have Dr. Howes's book in print again, and its reissue will enable beekeepers of a new generation to have copies of their own. If they know it at all, it is likely to be a copy from a library, or one lent unwillingly by its owner—it is such a useful book to refer to that no one would care to lend it for long. It is one of the very few books of which I need three copies; one at IBRA, one at home, and one in County Kerry in Ireland where I spend some weeks each year.

Since 1945, when the book was first published, much research work has been done on plants from which bees obtain nectar, pollen, honeydew and propolis. The journal *Apicultural Abstracts*, started in 1950, has reported about 2,000 new publications on them, out of a total of 20,000 on all subjects to do with bees and beekeeping. And yet Dr. Howes's book is still valid: his principles are fundamental ones, and the opening sentence of each section of the book is as proper today as when it was first written. It is one of the really living books to do with bees. In the balance between the different plants, however, today's situation is significantly different from that in 1945. Changed agricultural practices have reduced the importance of clover and of weeds and hedgerow plants, but have introduced new sources of honey, notably oil seed rape which is the subject of an extra section written for this edition by Dr J. B. Free of Rothamsted Experimental Station. Meanwhile, taxonomists have revised the names of some of the plants, which have been updated accordingly, with cross-references where necessary.

Many characteristics of bee plants that were only in their infancy in the 1940s have now become subjects of quite extensive study, and the literature on British bee plants has been enriched by many books and articles from Continental Europe and elsewhere. One of the greatest changes in the last thirty years has in fact been the interaction and co-operation between apicultural science in the different countries of the world. The Supplementary Bibliography on pages 221 to 223 of the present book provides a key to publications likely to be especially useful to readers.

The physiology of nectar secretion is much more completely understood than thirty years ago, and also the intricate sequence of interactions that leads to the making of honeydew honey by the bees: interactions between plant host and honeydew-producing insects, and between these insects and ants, and the microscopic constituents of honeydew honey that enables us to identify it as such. Correspondingly, the study of microscopic constituents of floral honey, known earlier as pollen analysis and now as melissopalynology, has become a subject in its own right. Chemical studies of propolis have yielded knowledge of a series of components with interesting properties; pollens of many plants have been analysed chemically, and found to have an impressive nutritive value. Honey has received even more attention than any of the above subjects, and a book published in 1975 (reference 33) cites 1,400 publications, many post-1945. Plant-insect relationships in pollination, and the special importance of the hive bee, are subjects of even greater economic importance now than when Dr. Howes wrote about them.

The idea of *growing* plants for bees is represented in nearly half of Dr. Howes's introductory sections, and this idea has taken on a new significance in the past few years, when the concept of self-sufficiency has become a popular aim. The book was written in a period of bee-keeping expansion, as the opening to Dr Howes's Preface shows. That phase was followed by a recession, but we are once again in a period of expansion, and today's new beekeepers differ from those of the 1940s in that they are in general better educated, with more scientific knowledge and a greater readiness to use reference books of which Dr Howes's is such an excellent example. I hope and believe that it will serve them as well as its predecessor served the last generation.

IBRA,
Hill House,
Gerrards Cross,
Bucks. S79 0NR.

Section 1

Plants and the Beekeeper

NECTAR AND NECTAR SECRETION

The production of nectar by plants is all important to the beekeeper for without adequate sources and supplies of nectar he cannot obtain honey (34). While a certain amount of useful knowledge has been acquired, particularly in recent years, with regard to nectar and the best conditions for its free production in some of the major honey plants, there is still much that is shrouded in mystery and has yet to be unravelled. This applies not only to most of the minor honey plants but to many of the major ones as well. Why is it for instance that heather is so capricious as a nectar yielder and why is hawthorn so fickle, yielding a good honey crop one year but nothing for many succeeding years, even under apparently favourable weather conditions? (33)

The fresh nectar of flowers is little more than a weak solution of cane sugar, generally containing 60 to 90 per cent water, with small quantities of other substances such as essential or volatile oils, flavourings, gums and traces of mineral matter. These lesser ingredients become of importance as the water is driven off the nectar in the hive and the nectar is converted into honey, for they determine the aroma and flavour of the honey and account for the well-known characteristics of honeys from different floral sources. Nectar is usually colourless but may be slightly coloured. It is not unusual for it to be noticeably scented or to possess a spicy taste. In some flowers it is produced very copiously as in the honeysuckle, certain orchids, the tulip-tree (*Liriodendron*) in North America and some of the Cape proteas or sugar bushes. The latter are so rich in nectar that it was a common practice in early days at the Cape—before the advent of cheap sugar—to boil the flowers in water and to concentrate the liquid to the consistency of a thick syrup for table use. On the other hand many flowers, often of great importance to the beekeeper, secrete only very small quantities of nectar. But the honey bee, with her specialized tongue or sucking mechanism, is able to absorb the most minute quantities of nectar—even when the surface appears only moist (59, 60).

The function of nectar in the flower is to attract insects (or in some cases birds) and so assist in bringing about cross pollination. While some insects, such as beetles, will visit flowers only for the pollen which they eat on the spot, quite a number of other insects like flies are attracted only by the nectar and do not eat or make use of the pollen in any way. Thus the presence of nectar as well as pollen is likely to lead to an increased number of visitors.

The sugar concentration of nectar in the flower and the changes that take place in it during the day have now been studied in the case of many plants. This applies particularly to tree fruits, clovers and other crops where studies in pollination with a view to increased fruit or seed setting have automatically led to investigations on the sugar content of the nectar. The concentration of nectar in the early morning is often very low, sometimes too low to attract the honey bee, but as the day advances various factors such as sunshine, wind, a rising temperature or reduced humidity assist in concentrating the nectar so that by midday the sugar percentage may be doubled or even three or four times as much. With some tree fruits such as the apple, it is not unusual for bees in the early part of the day to work the flowers only for pollen, but to turn their attention to nectar later when it has become more concentrated and to them more palatable. It is due to such differences in the sugar concentration of nectar that bees will forsake one kind of plant for another as the day advances. In the case of an open type of flower evaporation is likely to be more rapid and the sugar concentration increases at a quicker rate than might apply with a more closed type of flower. In fruit orchards in Surrey for instance it has been noticed that bees will forsake gooseberry and currant flowers for cherry blossom. The fact that they have been known to favour dandelions in preference to plums, and charlock to other fruits in some districts is probably also due to differences in the sugar concentration of the nectar.

Numerous devices, in some cases very ingenious and efficient, are to be found among different kinds of flowers for protecting the nectar from rain or from short-tongued insects that are unlikely to be effective in pollination. This often takes the form of dense hairs immediately above the nectar and attached to the side of the flower tube or to the bases of the stamens. On the other hand some flowers, such as the maples and the bulk of the carrot family (*Umbelliferae*) offer little or no protection to the nectar which is entirely exposed and affords a free feast to all. Flies and other short-tongued insects are frequent visitors to such flowers.

Many factors are known to affect nectar secretion in plants. Some of the more obvious and important of these are temperature, humidity, nature of the soil, soil moisture, wind and the age or vigour of the plant. With different plants these factors vary in importance and it is this that makes it difficult to assess the optimum conditions for a good nectar flow with any particular nectar plant. For instance soil is of primary importance in the case of heather and the clovers, the one requiring an acid and the others an alkaline soil. But with willow-herb and blackberry, both good nectar sources, the nature of the soil seems to be of little consequence for both grow freely and are much worked by bees for nectar in all types of soil, whether light or heavy, acid or chalk.

Some plants, such as willows and white clover, will secrete nectar at quite low temperatures, temperatures probably lower than those normally required for bees to fly. On the other hand there are plants such as the false acacia (*Robinia*) that will not secrete nectar unless the temperature is fairly high. This is the reason why the false acacia is a good honey plant in regions of hot Continental summers, whereas with the average cool English summer it is of little account as a source of nectar. Other trees such as the tulip-tree (*Liriondendron*) and tupelo (*Nyssa*) renowned as nectar yielders in their native land, North America, seem to be of little or no value for nectar in Britain, possibly for similar reasons.

With some important honey plants the opinion is held that cool nights followed by fine hot days provide the best conditions for a free nectar flow. This doubtless may hold with many plants, including white clover, and may explain why the best honey districts in many countries are to be found at the higher altitudes, the nights being cooler there. However, it certainly does not appear to hold with the lime, where warm nights followed by warm sultry days, but not fine hot days, give the best nectar flow and honey crops (see Lime), provided always, of course, there is adequate soil moisture.

Winds may be favourable or unfavourable according to circumstances. Severe cold winds are always likely to be inimical, but mild drying winds after a period of rain can have a distinctly beneficial effect.

The state of health, age, or vigour of the plant may be important with regard to nectar secretion. With heather it has been found that nectar is produced more copiously from comparatively young plants, i.e. heather recently burned over, than by those that are tall and woody and many years old. If growth is impeded through inadequate soil moisture and lack of rain, or there is any suggestion of wilting, the production of nectar ceases immediately. Drought may often be the cause of a dearth of nectar and a poor honey season, particularly with shallow-surface-rooting plants. In the case of white clover it is responsible for failure in some seasons in the south of England. With nectar-yielding trees like the lime and other deep-rooting plants drought is not so likely to be serious.

It is in this connection, i.e. the establishment or maintenance of adequate soil or subsoil moisture, that the weather some time before the honey flow, even months ahead, may be important. In some districts it is held that plenty of rain in the spring is necessary for a good nectar flow the following summer.

Sunlight is believed to have a bearing on the stimulation of nectar secretion with many plants, apart from the question of temperature. It has frequently been observed for instance that when some plants are shaded and others of the same kind are in full sun, the former may be completely ignored by bees, but the latter visited freely. This has been observed by the writer to hold very forcibly in the case of heaths at Kew, while the same has been recorded by other observers with other plants, including clovers. Statistics show that good, medium and poor honey seasons in the British Isles over a long period of years work out in the ratio of 1:2:1, or roughly in every eight-year cycle two will be good honey years, two will be poor and four will be average. The same ratio has been recorded for parts of Europe, e.g. Switzerland. For every district there is a normal average honey crop, dependent upon the amount and the nature of the bee forage available. Some years it will rise above the average. In others it will sink well below it. Abnormal circumstances such as the extensive ploughing up of clover pastures for cereals and root crops, as takes place in wartime, may of course fundamentally alter the honey-producing capacity of any district.

Locality is of prime importance in beekeeping. Not only does it govern the method of management adopted but a distance of a few miles is sometimes sufficient to account for a marked difference in the honey-producing capacity of a district. Often this is due to differences in the soil, which may have a direct bearing on nectar secretion. Topography may also be important, for some plants, such as white clover, are believed to secrete better on hilly or sloping than on flat land (14). This may be mainly a matter of drainage and soil aeration. With heather some consider plants on peat and bog land do not yield much nectar whereas those on hill land with granite or ironstone subsoil yield well (22). On the other hand in dry seasons vegetation at the lower levels may suffer less from drought than that on exposed hillsides and so be in a better condition for nectar secretion. In the case of early- and late-flowering nectar plants topography may be important with regard to the existence of frost pockets and cold air drainage, for flowers injured or destroyed by frost are of no use for nectar.

Nectaries

Nectar is produced in the flower by special organs, called nectaries. Often these consist of little other than small groups of specialized secretory cells, but in some plants they may be more elaborate. Among the plants that are valuable to the hive bee for nectar an interesting range of different types of nectary is to be found.

The usual place for the nectary or nectaries is at the base of the flower, where it frequently takes the form of a raised ring or ridge of secretory tissue, often yellow in colour. The nectary may, however, be situated on almost any part of the flower—sepal, petal, stamen, or pistil—but these positions are less common.

The lime affords a good example of nectar being secreted by the sepals. These are boat-shaped and lined with hairs which assist in holding the drop of nectar as it forms in the hollow or inner surface of the sepal, for the flower is generally in a pendent position. The nectaries themselves, slits of secretory tissue, are invisible to the naked eye and hidden by the hairs. In many of the *Cruciferae*, the family which includes charlock and mustard, the sepals are bent back or curved, and serve as containers for the nectar. It is not secreted by them, as in the lime, but by nectaries at the base of the flower.

Instances of nectar being secreted by the petals are afforded by the snowdrop and the tulip-tree. In the snowdrop the nectar is formed in grooves on the inner surface of the petals and in the tulip-tree on the yellow patch near the base of each petal. Interesting cup-like nectaries can be found in the Christmas rose (*Helleborus niger*) and winter aconite (*Eranthis*) where the whole petal has been modified into a vase-like structure to secrete and hold the nectar. In many flowers there is a spur or pouch for the nectar, the garden nasturtium and some of the shrubby honeysuckles being good examples.

Instances of stamens fulfilling the role of nectar secretion are to be found in the wild clematis and the violet. In the former the nectar appears as tiny droplets on the filaments or stalks of the stamens. In the violet the enlarged appendages of two of the anthers bear the secretory tissue producing the nectar. This collects in the spur of the petals immediately beneath it.

There are many examples of nectar being secreted by nectaries on the pistil or ovary of the flower. In the garden hyacinth three small dots may be seen near the apex of the ovary. These are the actual nectaries and frequently beads of nectar may be seen adhering to them. Many other allied plants secrete nectar in a similar fashion. The scabious and the teasel (in fact all the family *Dipsaceae*) also secrete nectar from the ovary. In the marsh marigold (*Caltha*) the nectaries take the form of conspicuous appendages attached to the female organs or carpels.

There is much variation in the shape and form of the nectary itself. In the almond, plum and many other rosaceous plants, the receptacle or base of the flower is hollow and the nectar may appear first as tiny dots or droplets anywhere on its surface. These gradually enlarge and may eventually coalesce. In the thistles and other members of the same family (*Compositae*) a small ring at the base of the flower tube secretes the nectar. As the flower-tube is usually narrow and partly occupied by the pistil, the nectar soon rises in the tube, sometimes to a height of several millimetres. Much the same holds with the clovers, but in this case it is the staminal tube and not the corolla tube that holds the nectar. In the willow catkin the individual flowers are much reduced, consisting of only a single scale plus one or two stamens, or pistil in the case of the female flower. The nectary takes the form of a knob at the base of the scale which produces and becomes surrounded by a single drop of nectar. The nectar is usually easily seen in willow catkins, male or female, that have been kept in a warm room overnight (40, 51, 52).

HONEY IN RELATION TO NECTAR SOURCE

The nature and quality of honey is mainly governed by the plant source from which it is obtained. This determines also the flavour, aroma, density, colour and brightness of the honey (33). In many countries large quantities of honey are obtained from a single plant source, as for instance in the extensive buckwheat areas in Russia and the eastern United States, citrus in California and Palestine, eucalyptus in Australia and tea-tree honey in New Zealand. In Britain (35), however, with the exception of heather honey, most of the honey produced is a blend of some sort or other, although clover may predominate. Some (14) consider the fact that English honey is for the most part a natural blend accounts for its popularity as a table delicacy over most imported honeys, and that a single plant honey is unlikely to be so attractive to the palate as a natural blend prepared by the bees.

The main honeys that are produced in the British Isles in any quantity in a reasonably pure or only slightly blended state besides white clover and heather are lime, sainfoin, fruit blossom, mustard or charlock, hawthorn and perhaps buckwheat, sycamore or dandelion in restricted areas.

The term 'flower honey' sometimes used among beekeepers is intended to denote any honey other than heather. Another term often heard is 'tree honey'. This is applied to the early season honey obtained mainly in the south from fruit blossom and any other early-flowering, nectar-yielding trees that may be in flower at about that time, such as sycamore, the wild cherries and perhaps horse-chestnut, hawthorn or holly. 'Tree honey' is obtained from different plant sources in varying amounts according to locality and is itself therefore very variable, but is generally rather dark and strong fla-voured. Many of the dark or unusual honeys seen at honey shows come in this category (14). The term 'tree honey' is not applied to lime.

There is a fairly wide colour range among English honeys although the majority are some shade of amber. Very pale or water white honey has been obtained from willow-herb (*Epilobium*). Lime is often distinctly greenish, bell heath (*Erica cinerea*) reddish, and heather and buckwheat very dark. There is no direct correlation between flavour and colour but in general the lighter or pale-coloured honeys are milder in flavour than the dark coloured.

It is generally believed that the type of soil or subsoil on which honey plants grow has a bearing on the resulting honey and that in general clay soils give a darker honey than is obtained from the same plant on light or sandy soil. The presence or the relative amounts of certain elements such as iron, manganese and copper are also believed to influence the colour of the honey (30).

Vitamins have now been shown to be present in various honeys but only in very small amounts (from 0 to 20 mg. per 100 g.), the quantity being insufficient seriously to affect or enhance the protective food value of the honey. It has been shown that honey from different floral sources shows much variation with regard to vitamin content and it is thought this may be dependent upon the amount of pollen actually present in the honey.

While honey production takes place in the British Isles under all sorts of conditions some of the best honey-producing districts are those on the plains overlying the chalk in the south and east of England, and in sheep areas where clovers are generally an important constituent of the vegetation. It is a common saying that 'bees follow the sheep' (29). Most of the larger or commercial honey producers in the country have their apiaries or 'outfits' in areas where the soil has a high lime content, i.e. on the chalk belt which runs from Dorset through the Chilterns and on into Norfolk. Similar conditions exist in the Cotswolds and elsewhere.

Among the more distinctive types of English honey, that from white clover is one of the most esteemed and has a more or less universal appeal. It has a delicate flavour and aroma and good density. The colour is light, varying from water white to pale amber. This may be influenced by the nature of the soil but it is also known that when the flow is a rapid one and clover nectar available in abundance the honey is likely to be lighter in colour than when the flow is not so fast and more protracted owing to less suitable weather conditions. Another of the virtues of clover honey is that it crystallizes or 'sets' with a smooth, fine grain.

Lime honey in a more or less pure state is available in some localities, mainly urban areas. The density is not so good as that of clover and the colour slightly greenish as a rule. The flavour is distinctive and suggestive of peppermint. In some seasons the honey is darkened and spoiled by the presence of honeydew, lime trees being bad offenders in this respect.

The honey from heather or ling is very distinctive and quite different from all other English honeys. Its thick, jelly-like consistency is perhaps its main feature. This prevents its being extracted with an ordinary rotary extractor and presses are used to obtain run honey. The air bubbles that form on pressing remain in the honey thereby imparting a characteristic appearance, and do not rise to the surface or disappear as in other honeys. In spite of its dark colour and strong flavour heather honey is much sought after and always commands higher prices than other honey. Bell heath (*Erica cinerea*) or bell heather as it is also called yields quite a different type of honey which is not gelatinous and may be extracted in the ordinary way. It is reddish in colour (port wine) when pure with a pronounced flavour somewhat resembling that of ling. Although this heath is usually to be found growing with ling, honey from it is sometimes obtained in a reasonably pure state for it is in flower a good deal earlier than ling.

Honey from the blossoms of tree fruits (apple, pear, cherry, or plum) is variable according to the kind of fruit grown but is usually rather dark and strong in flavour. Honey from miscellaneous fruit blossoms probably constitutes the bulk of the so-called 'tree honey' already described. Where extensive apple orchards exist a reasonably pure apple honey might be obtained in favourable seasons. Apple honey may be either light or dark amber in colour and fairly thick. The flavour is strong at first but improves with age.

Sainfoin yields one of the most distinctive honeys that are obtained in Britain. This clover-like plant is usually grown only in chalk districts in the south and is in flower in the latter part of May and in June. The honey from it is deep yellow in colour, with a characteristic flavour and aroma. Although bright and sparkling in appearance the density is not so good as that of white clover. Sainfoin section honey is equally distinctive.

Honey from charlock and mustard, often obtained in agricultural districts, is notable for the rapidity with which it granulates, sometimes within a few days of being extracted. It is a good quality honey and light in colour. The flavour is inclined to be strong at first, even slightly pungent, but this passes off with age.

Field or horse beans are another not uncommon source of honey in farming districts, the flowers being available to bees fairly early in the season. Honey from them has a pleasant mild flavour but granulates fairly quickly with a coarse grain.

Another farm crop yielding a distinctive type of honey is buckwheat. However, it is only grown to a limited extent and in certain districts, usually as a catch crop. Honey from the flowers of buckwheat is always dark with a strong flavour and is not generally liked by those accustomed to mild flavoured honey.

Occasionally in the south of England there are instances of dandelion honey being obtained in what is considered to be a fairly pure state. It is pale or deep yellow in colour with a strong flavour not appreciated by everyone. It crystallizes fairly quickly with a coarse grain.

May or hawthorn is a good source of nectar in some seasons but not often. Honey from may blossom, when obtainable in a reasonably pure form, is of very good quality. It is generally rather dark and very thick with a rich appetizing flavour.

The sycamore is another tree that is sufficiently common in some districts to be a source of surplus honey in favourable seasons. Sycamore honey generally has a greenish tinge and the flavour is not of the best, especially when fresh. It granulates slowly with a coarse grain.

Blackberry or bramble is one of the commonest of British plants and at the same time one of the most useful to the beekeeper. In some districts, where clover and lime are absent, beekeepers consider that what honey they may get before the heather flow is mainly blackberry. Honey from blackberry is dense and slow to granulate but the flavour is not of the best in the opinion of many.

Another wild plant which is sometimes very prevalent, and, like the blackberry, not fastidious as to soil or situation, is the willow-herb or fireweed. Where fires have occurred or extensive areas of woodland been cleared it may be very prevalent for a time, brightening the landscape with its large pink flowers, and constituting a useful source of late season honey. The characteristic feature of the honey is its pale colour, often water white. It has not a pronounced flavour and is useful for mixing with other strong flavoured honey to tone it down.

Among the imported honeys sold in Britain which may have a characteristic flavour due to a particular floral source are the following: *orange*—California, Syria and Palestine; *logwood*—Jamaica; *clover*—Canada, U.S.A. and New Zealand; *eucalyptus*—Australia; *buckwheat*—Russia; *thyme*—Hymettus, Greece and Syria; *lavender*—Syria; *rosemary*—Narbonne, France; *peach*—Italy; *wild acacia*—Syria. In some instances these honeys are sold more or less as luxury lines in high-class food shops, having been specially imported.

NOTES ON UNPALATABLE AND POISONOUS HONEY

Honey that is poisonous or harmful to human beings has been recorded from many countries and since classical times, the earliest account being that given by Xenophon during the memorable retreat of the Ten Thousand in the year 40 BC, the source of the poisonous honey being considered to be the pontic rhododendron (*R. ponticum*) (49). The deleterious properties of poisonous honey are believed to be due to toxicity of some kind in the nectar itself from which the honey is prepared (64). This toxicity is of an elusive or fugitive nature and in the majority of cases disappears as the honey ages or ripens. Usually such honey is only harmful when in the raw or uncapped state and as soon as it is capped over by the bees it becomes safe to eat.

In the British Isles poisonous honey is almost unknown. The word 'almost' must be used for there are one or two instances where the common pontic rhododendron may have been responsible for honey found to possess harmful properties. Normally this plant is not of much consequence as a source of nectar for hive bees but it would appear that in some seasons a certain amount of nectar is obtained from it. A sample of honey from Cobham, Kent, alleged to be from this rhododendron, was found to possess emetic properties by all who sampled it (9). In another case, at Camberley, Surrey, where new comb honey was eaten for breakfast the sickness and symptoms experienced (giddiness, distorted vision, perspiration, etc.) were very similar to those recorded for rhododendron poisoning in eastern Europe, a good account of which has been given by Mosolevsky (*Bee World*, 1929, 141; 1942, 31). In this instance some, but not all the members of the family who had partaken of the honey for breakfast were affected, which suggests that possibly only some cells of the comb contained the poisonous honey. Numerous rhododendrons and some azaleas grew in the vicinity. What may be another case of poisoning in England is that recorded by a Nottinghamshire doctor who had known boys who had robbed bumble bees' nests to suffer from vomiting, purging and abdominal pains (*Scottish Beekeeper*, Feb.–March, 1942). Bumble bees are better able to procure the nectar from rhododendron flowers than are hive bees, on account of their longer tongues, and visit them more freely. Poisoning from rhododendron honey is said to be much in evidence in some districts in the Caucasus as soon as the consumption of comb honey commences. It occurs to some extent every year and is more pronounced in dry than in wet seasons. The first symptoms arise some three or four hours after eating. Fatal cases with it occur mostly with children (Mosolevsky).

Plants known to yield poisonous or unwholesome honey in other countries include the following: South Eastern Europe—*Rhododendron ponticum*; Japan—*Tripetaleia paniculata*; New Zealand—*Melicope ternata*; South Africa—*Euphorbia spp.*; North America—*Kalmia latifolia, Gelsemium sempervirens* and possibly species of *Pieris, Andromeda* and *Leucothoe*. It is interesting to observe that these plants, with but three exceptions, are members of the heath family (*Ericaceae*). However, fortunately for British beekeepers, not all of the heath family produce honey that is liable to prove unwholesome, for that of heather or ling and of bell heath is well beyond suspicion.

Among the above *Kalmia latifolia* (calico bush or mountain laurel) is sometimes cultivated as an ornamental shrub in the British Isles, and is a large evergreen, rather like a rhododendron, with clusters of white or pink flowers at the ends of the branches. However, it is probably nowhere sufficiently common to alarm the beekeeper. In regions where the plant is common in the eastern United States, farmers and beekeepers have been known to first feed honey which they look upon with suspicion to the dog. If no ill effects are noticed within a few hours the children and other members of the household are allowed to have it (20). Species of *Pieris, Andromeda* and *Leucothoe* are also sometimes cultivated in Britain but not to any extent. The Japanese shrub, *Tripetaleia paniculata*, which has been cultivated at Kew, is the source of unwholesome honey in mountain districts in Japan where it occurs in abundance. The honey is said to have a pungent taste and the severity of the poisoning to vary with different individuals. Recovery usually takes place within several hours or a few days (*Bee World*, 1925, 4–5). In South Africa noors honey, obtained from euphorbias, the dominant vegetation in some areas, produces a hot burning sensation in the mouth and throat which is increased rather than decreased by drinking water. Some species of eucalyptus in Australia yield honey which many regard as unpleasantly strongly flavoured, whereas other eucalyptus species yield honey of the finest quality.

The main sources of objectionable or ill-tasting honey in the British Isles are ragwort and privet. The honey from both these plants will spoil other honey if it becomes mixed with it to any extent. Fortunately neither is very prevalent in most districts and they are not usually considered to be a serious nuisance. Both flower fairly late in the season, the main flowering often taking place after the honey crop has been removed, in which case their presence is welcomed by the beekeeper as affording a useful late source of nectar and food for the bees themselves. There is nothing to suggest that the bees find the honey in any way objectionable.

Ragwort honey is a deep yellow in colour and possesses a strong flavour and aroma, somewhat nauseous in fact. This weed is most common on poor and neglected pastures but may be found in all sorts of situations and be much more prevalent in some years than in others. It is very common on the Breckland. Ragwort withstands drought much better than clover and in dry summers when clover fails or is poor as a nectar source, bees are prone to work ragwort more intensively and ragwort honey is more noticeable.

Privet is perhaps more the concern of the urban than the rural beekeeper. Overgrown or neglected privet hedges produce an abundance of blossom rich in nectar. Bees are quick to make use of it but usually the quantity available is not sufficient to affect the main crop honey. Privet honey is dark in colour, fairly thick and distinctly bitter in taste, quite uneatable in the opinion of many. It is this bitterness which is liable to spoil good honey.

The fact that bees do themselves suffer ill effects or poisoning from the nectar of certain plants is well established. One of the most outstanding examples is the 'buckeye poisoning' which is troublesome to beekeepers in the southern United States and is due to the nectar collected from buckeye blossoms, trees or shrubs very similar and closely related to the common horse-chestnut. In buckeye poisoning it is the brood or young bees rather than the adult bees that are affected. Loco weed (*Astragalus spp.*) is also held responsible for the poisoning of bees (adult bees) in North America. In some European countries conifer honeydew is believed to be a cause or one of the causes of bee paralysis. Honeydew from limes has also been shown by experiment to be harmful to bees but it is not known to what extent it may poison bees under natural conditions. Honeydew is present on limes in Britain in most seasons, particularly in hot dry summers, but bees do not always collect it.

The nectar of some of the late flowering limes appears to be inimical to bees, at least in some seasons. It is a common sight to see dead or dying bees, mainly bumble bees, underneath trees of *Tilia petiolaris* and *T. orbicularis* when they are in flower. The degree of poisoning varies from year to year. In some seasons the number of dead or dying bees is very large, even to the extent of subsequently enriching the ground as the following observation indicates: 'In 1908 the bodies of innumerable bees, poisoned by the flowers of *T. petiolaris* at Tortworth, had so much manured the ground under its outer branches, that a very green ring of turf was visible in the autumn following, and was noticed by the Earl of Ducie to be even more conspicuous in 1909' (*Trees and Shrubs of Great Britain*, Elwes and Henry). The writer has found that bees, both bumble and hive bees, picked up from the ground under these trees, frequently recover. However, while in this stupefied condition they are doubtless easy prey to various natural enemies as the mutilated carcasses of many bees suggest—possibly the work of tits.

It is quite possible that other plants in the country visited by bees may have a similar if less pronounced effect, but little or nothing is at present known in regard to this. There are indications that poppies may also have some sort of stupefying effect on bees.

POLLEN

Pollen is of vital importance to the honey bee and so to the beekeeper, for it is the only source of nitrogenous food of bee larvae, the amount of protein in ordinary honey being negligible, usually about 0·2 per cent (62). Without it they cannot grow and develop. The absence of pollen would therefore soon lead to the extinction of the colony (31).

Fortunately in the British Isles there is usually adequate pollen available from wild or natural sources during the breeding months of the year. This is not the case in many other countries, such as parts of Australia and the southern United States, where acute pollen shortages occur at certain periods of the year when bees are normally active and breeding. Such dearth periods for pollen often constitute a serious and difficult problem for beekeepers in those areas.

Every district or locality has its own pollen cycle for the year and this remains much the same year after year, although some seasons commence early and others late. The pollen cycle for most British beekeepers commences in early spring with such plants as coltsfoot, hazel, crocus, gorse, willow, etc. Then follow other well-known pollen sources like the dandelion, tree fruits, hawthorn, and horse-chestnut, along with innumerable other spring and summer flowers. These are succeeded by the clovers, limes and a host of wild and cultivated late summer plants. The pollen cycle terminates with such subjects as heather, thistles, ivy and in the case of urban beekeepers autumn garden flowers such as Michaelmas daisies, golden rod, sunflowers, etc. In the British Isles the areas most likely to suffer from pollen shortage in the early part of the year are the open flat districts where there is little or no woodland and streams harbouring willows are few and far between. Gorse is a useful standby in many districts, notably heath areas, especially as it is in flower more or less through-out the year.

The total quantity of pollen carried into the average hive in a season is now known to be very large, probably much larger than is generally supposed (43). It has been calculated that about ten average bee loads of pollen are necessary to produce one worker bee and that on an average 1 lb. of pollen rears 4,540 bees, which works out at about 44 lb. of pollen for an average colony's breeding require-ments in a season (44). As much as 71 lb. of pollen has been obtained from a single colony in a season by means of pollen traps. (*The Role of Pollen in the Economy of the Hive* by E. Todd and R. K. Bishop, U.S. Bureau of Entomology, 1941.)

Bees may visit flowers for the express purpose of collecting pollen or, as more generally happens, pollen is collected as a side-line during the collection of nectar, for pollen sticks readily to the hairy body of the bee. The manner in which a bee uses its legs for cleaning or partially cleaning the different parts of its body of pollen and packs it into the pollen baskets of the hind legs, generally while in flight, is well described in several of the standard works on bee-keeping.

When visiting flowers for pollen the bee varies the method of obtaining it with different flowers, according to their structure. In the open type of flower like the apple, hawthorn or blackberry the bee draws the anthers towards it with its forelegs as it runs rapidly over the flower, sometimes biting the anthers with its mandibles, and re-moves the pollen that adheres to its body to the pollen baskets at intervals. In dealing with the flowers of a catkin such as hazel or willow, the bee may run some distance up the catkin, then fly away a short distance to pack the pollen, returning to repeat the process. sometimes it does not even alight on the catkin but brushes the anthers while suspended in flight in the air. When the bee is working the tubular type of flower, as in many bulbous plants and in the mint family, or the closed type of flower as in the clovers, the pollen adheres mainly to the mouth parts or forelegs and is removed and packed into the pollen baskets as the bee flies from flower to flower (18).

Most plants yield pollen throughout the day under favourable weather conditions, but there are some, such as many roses, grasses and maize or sweet corn, that yield pollen only in the morning, the anthers ceasing to dehisce by midday.

The chemical composition of pollen from different plants is known to vary considerably. Some pollens like those of sainfoin and the dandelion are much more oily than others as anyone who has examined pollens microscopically will know. There is a wide range of variation in oil and fat content of different pollens, as well as in other constituents such as protein, starch or carbohydrate, vitamins and mineral matter (18).

It is well known by beekeepers that bees may show a marked preference for one kind of pollen over another which may be equally abundant and equally easily obtained. To what extent this may be correlated with the chemical composition and the believed nutritive value has yet to be ascertained, at least with the great majority of plants. As bee food it is the quantity or percentage of digestible albumen or protein in the pollen that matters. This was found by one investigator to vary from 10 per cent in the case of fir to 46 per cent (dry weight) in the case of hazel. Furthermore it decreased with age, that of hazel becoming 18 per cent after one year and 14 per cent after two years (*Bee World*, November 1940). This shows that pollen collected and stored dry to be subsequently fed to bees, as can easily be done, is likely to be of little use, for it will have lost much of its food value. The vitamin value of pollen also diminishes on dry storage. When more is known of the nutritive value of different pollens for bees it should be possible to state with more certainty which plants are most worth while growing by beekeepers as early sources of pollen.

Some kinds of pollen, or pollen under certain conditions, may be harmful to bees. That of buttercups (*Ranunculus*) is known to cause a type of poisoning in some countries (see Buttercup). Harmful results may also arise from the collection of pollen that has been frosted.

The predominating colour of pollen is yellow or cream, in numerous different shades, but many other colours are to be found. The colour of the pollen in the bee's pollen baskets, as it is carried into the hive, is invariably a source of interest to beekeepers. Those with hives near horse-chestnuts cannot fail to notice the brick-red pollen loads that are so conspicuous when these trees are in flower. Other plants with unusually-coloured pollen or which give rise to unusually-coloured pollen loads include:—meadowsweet; loosestrife (green or greenish); poppy (black or purplish); heather (slate grey); phacelia and scilla (blue); sainfoin (yellowish brown); willow-herb (bluish green); sheep's scabious (mauve); blackberry and raspberry (white); and purple deadnettle (bright orange).

There are many conflicting descriptions of the colour of different pollens to be found in bee literature. This is probably due to the fact that the colour or shade of a pollen may vary with age and its appearance in the bee's pollen baskets after it has been moistened with nectar or honey and patted down may differ from its colour in the fresh state as it appears dry on the newly-opened anther of the flower. Whether viewed in a thin layer or small quantities is also important, as is the background. When it is examined microscopically the use of reflected or transmitted light, as well as the source of light, may influence colour. Furthermore, it is necessary to bear in mind that different persons do not always see colours alike. The use of standard colour charts as now often used in describing the colours of flowers and in scientific work generally, should afford the best means of describing or recording pollen colour. A glaring case of inaccuracy with regard to pollen colour is to be found in a well-known English bee book where the pollen of the pear is described as red, the writer having obviously mistaken the colour of the anther itself for that of the pollen.

Pollen may be responsible for a characteristic colour in beeswax and in the actual honeycomb and cappings. When wax is first secreted as small scales on the abdomen of the bee it is invariably colourless, whatever the source of food of the bee. It is afterwards, when the scales are manipulated and made into comb by the bees and with the passing of time and the virgin comb being worked over or polished by young bees, that a characteristic colour may develop. This is believed to be mainly due to the wax absorbing colouring matter from the pollen that bees are bringing into the hive at the time. Comb made from old wax is darker in colour than virgin wax.

Many pollens are rich in oil and wax-soluble substances, usually of some shade of yellow or orange. Even if the pollen is not in direct contact with the comb, as in honey supers, the colouring matter may reach it by way of the bees' feet, legs and body and by adhering to particles of propolis which stick to the bees' feet. The bright yellow of wax and comb produced when sainfoin is worked is familiar to many beekeepers, also the staining of the comb frames and woodwork of the hive that takes place with it. It has been shown by experiment that while some pollens release their colouring matter very freely to wax others do not do so at all (*Bee World*, 1935, 117). Imported beeswax shows a much wider range of colour than that produced in the British Isles, reddish, grey, or greenish shades being sometimes met with. Impurities or adulteration may also influence the colour of crude commercial beeswax.

Pollen grains are minute and produced in great abundance in many flowers, a good example being the single peony where it has been estimated that one flower may produce over three and a half million pollen grains (20). In the case of a single catkin of the birch as many as ten million pollen grains have been estimated to be present (6). Although so small and far too minute for observation with the naked eye, or even a hand lens, pollen grains might be said to constitute a world of their own in much the same way as do bacteria, diatoms, etc. Viewed microscopically they show endless differences in form and structure, in the moulding or sculpturing of the surface, the presence or absence of spines or other outgrowths and in the number of pores or grooves. In some the surface is quite smooth, and in others striated, reticulated or cellular.

There are pronounced differences in size among the pollen grains of different plants. Their diameters may vary from less than 10 microns (1 micron = 1/1,000,000 metre) in the case of such plants as forget-me-not (*Myosotis*) and goat's rue (*Galega*) to over 100 microns (or even 150) in the case of crocus, evening primrose, hollyhock, mallow, vegetable marrow and other gourd plants.

In some plants or groups of plants the pollen grains are distinctive, for instance those of the heaths and many other members of the heath family (azaleas, rhododendrons, etc.) are in tetrads or gronps of four, arranged like one orange placed on three other oranges. In the forget-me-not family (*Boraginaceae*) which includes many good bee plants like borage, bugloss, etc., they are dumb-bell shaped. Quite frequently pollen grains are bound together or mixed up with strands of viscin, as in the willow-herb. This is a sticky substance and causes the pollen to cling more readily to insects and therefore to effect cross pollination. In contrast to those pollen grains of distinctive structure there are large numbers, in several families, that look very similar, being 30–40 microns in diameter, triangular with three grooves or pores. Such grains, when found in honey, are often extremely difficult to identify with any certainty. Those of the various clovers belong to this class.

The importance of pollen grains in honey and the evidence they may afford with regard to the botanical source of honey and to adulteration is well known. For instance, if honey alleged to be English is found to contain pollen grains of *Eucalyptus* or *Citrus* (orange) it has obviously been adulterated with imported honey, for these plants do not occur in Britain. A certain amount of caution has to be exercised in assuming that the presence of a large amount of a particular kind of pollen means that that plant is also the source of the honey. It is well known that honey may be moved from one part of the hive to another by bees. Honey might be stored in cells containing a certain amount of pollen collected some time previously, which of course becomes mixed with the honey. Furthermore, the pollen of some plants, owing to the structure or nature of the flower, is more liable to be present in abundance than that from other nectar plants. Honey from lime, willow-herb and fuchsia for instance contains, relatively speaking, few pollen grains (30). This is because the anthers are so placed that the pollen falls away from the nectar and is less prone to contaminate it than is the case with upright flowers. Heather honey may contain clover pollen although there may be no clover on the moors. This is because the supers used had previously been used for clover. There are also reasons why forget-me-not pollen may be unduly prevalent in honey (see forget-me-not) (60).

The microscopic study of pollen and the identification of pollen in honey is too large and involved a subject to be dealt with here. Those interested should consult the various textbooks that already exist in regard to it, e.g. those by Zander, Armbruster, Wodehouse, Hayes, Yate-Allen, and Erdtman. For publications on pollen in honey, see bibliographical references 33, 51 and 52. There is no reason why any beekeeper who possesses or has access to a compound microscope, and who has a good knowledge of the local flora, should not undertake a detailed study of the pollen his bees are likely to collect. Apart from the microscope, the equipment required need not be extensive or elaborate (8).

THE HIVE BEE AND POLLINATION

Although the honey bee is only one of many hundreds of different kinds of insect that may be responsible for pollination in fruit, farm or seed crops throughout the country, there are reasons why it is of special importance in this connection (32, 50). Its value as a pollinator is, in fact, considered to be far greater than its value as a honey producer (53). As a pollinator its importance was demonstrated in the years immediately following the first world war when bee diseases (notably acarine or 'Isle of Wight disease') had more or less decimated the country's bee population and fruit crops in particular were very poor.

Insects which are of importance as flower visitors include wasps, bees, and their allies (*Hymenoptera*), butterflies and moths (*Lepidoptera*), flies (*Diptera*) and beetles (*Coleoptera*) (36). Only a comparatively few beetles visit flowers or live on a floral diet. They are mainly interested in the pollen which they eat on the spot. Butterflies, moths and flies are attracted by the nectar alone. Thus it is only the bees that are regular collectors of nectar and pollen. They have their bodies specially adapted for the collection of pollen and are the only insects that feed their young with it. They are wholly dependent on flowers for their food, both for themselves and their young (61).

What makes the honey bee of greater value in pollination than other bees is the fact that it is available in fair numbers from early until late in the season. The whole colony hibernates and survives the winter whereas with bumble bees, wasps, etc., it is only the queen that survives the winter and it takes time for her to rear brood and establish a new colony in the spring. In some years, after severe or prolonged winters, the populations of wild bees and other insects may be very low in the spring. Many of the wild bees are active for only a short period of the year (thirty to ninety days), some early in the season others late, whereas hive bees are on the wing from early spring until late autumn.

The constancy of the honey bee, as compared with other insects, in working only one kind of flower on each trip from the hive is well known and was even referred to by Aristotle. This increases its value in the pollination of economic plants. A butterfly will flit gaily from one kind of flower to another, but a honey bee does not behave in this way. It is only when there is a scarcity of bee pasturage that the honey bee loses this habit of constancy. Under such conditions it is not uncommon to find its pollen loads consisting of very mixed pollen. Honey bees will, however, sometimes visit the flowers of closely allied plants on the same trip, even when there is an abundance of pollen. For instance, in the case of tree fruits they have been observed to visit the blossoms of plums after pears and cherries after apples. In working a head or bunch of flowers the hive bee works in a more thorough and methodical manner than other insects, including bumble bees. This is another of its virtues as a pollinator.

The value of bumble bees as pollinators, particularly for fruit, should not be overlooked. They are often very numerous, especially when there is plenty of waste or uncultivated land, woodland or hedgerows in the vicinity, for these are their natural breeding places. They possess one advantage over the honey bee as pollinators. It is that they will visit flowers under much colder and less favourable weather conditions than will the honey bee—essentially a fair weather insect that will not usually leave the hive at temperatures below 55 ° C. Quite frequently one sees these sturdy insects going about their business in rain or in cold wind when honey bees are confined to the hive. It has been shown, however, that flowers that have lost their petals through high winds or heavy rain are more likely to be visited by honey bees than bumble bees (Fox Wilson, *R.H.S. Journal*, 1926). While a colony of bumble bees consists of only some 50 to 200 individuals a hive may contain 30,000 or more honey bees.

The tendency in modern fruit culture is to have larger acreages under cultivation and fewer varieties than was the case formerly. The existence of fruit in large blocks generally means that the natural nesting places of wild bees are considerably reduced and their numbers become insufficient for effective pollination. It is then that the hive bee becomes of special value for it is the only pollinating insect that can be kept in a state of domestication and which can be artificially increased in numbers or moved from one place to another by man. Fruit trees in large orchards and in towns or built-up areas probably benefit most from the attentions of the hive bee for it is in such cases that the populations of wild insects are most likely to be insufficient. Where a few trees exist around homesteads on farms and in villages there are generally sufficient

B

wild pollinating insects for their requirements. The same applies with other plants. In a state of nature a correct balance between flowers and insects is to be found. It is when man upsets this balance by planting large orchards and fields that trouble starts and pollination difficulties arise.

In the more important fruit-growing districts of the country it has long been the practice to arrange for hives to be in the orchards at blossoming time to ensure adequate pollination—one of the first essentials of good crops. In the cherry orchards of Kent this is a regular practice with large growers, the value of sufficient honey-bee pollinators with this crop having been proved over and over again. It is sometimes noticeable in cherry orchards that trees nearest the hives carry much more fruit than those that are farthest away. In some cases trees on the outside of an orchard may carry more fruit than those inside, for they receive more attention from wild pollinating insects in the neighbourhood. At one time, in addition to local bees, skeps of bees were imported from the Continent, especially Holland, and placed in groups or at intervals in the cherry orchards. Bees are also brought from other counties to Kent for the purpose.

A large firm of fruit growers and jam manufacturers in Cambridgeshire maintains an apiarist to look after its hives, which are kept primarily for fruit pollination. In addition a number of their workers receive special instruction in dealing with swarms. Cider manufacturers and growers in the West Country make a special feature of apiculture in connection with their apple orchards for the same reason. A single apple blossom may produce from 70,000 to 100,000 pollen grains and it has been estimated that one honey bee may carry 50,000 to 75,000 pollen grains on its body, only ten of which are necessary for the complete fertilization of an apple flower. Abnormal or misshapen apples are often the result of incomplete pollination or fertilization of an apple flower. Some fruits, like strawberries, raspberries and blackberries, require several pollen grains for complete development whereas stone fruits, like the cherry and plum, require only a single functional pollen grain.

Different varieties of tree fruits vary considerably in their ability to set fruit with their own pollen. Some are entirely self-sterile, such as 'Cox's Orange' and 'Beauty of Bath' apples, while others like 'Bramley's Seedling' are self-fertile and do not require pollen from another source. Other varieties of apple are partially self-fertile. The same applies to commonly cultivated varieties of pear and plum and to cherries. With these self-sterile varieties cross pollination with another variety is essential for the production of fruit. It has been found that even the most self-fertile varieties produce much heavier crops when cross pollinated. This emphasizes both the need for mixed varieties in large orchards and the importance of the honey bee to the fruit grower.

In commercial fruit orchards it is usually considered that one strong hive per acre is sufficient for trees in full bearing and in the case of young trees one hive for 5 acres. It is essential that hives should be fairly strong, say 5 to 6 pounds of bees and 6 to 7 frames of brood. A certain number of bees is always necessary in a hive to maintain warmth, especially early in the season, and unless it is of fair strength there may not be sufficient surplus or flying bees for effective pollination. At fruit blossoming time in Britain the weather is prone to be cold, wet, or windy, sometimes for days on end, and too unfavourable for bees to fly. In some seasons there may be relatively few bright or warm days. When such days do arrive, therefore, it is desirable to get as much pollination effected as possible. The presence of sufficient populous stocks of bees is the best and possibly the only means of achieving this in a large orchard.

There is difference of opinion as to the best method of placing hives in an orchard for pollination purposes. Some favour spacing them singly at regular intervals, others prefer to place them in groups at wider intervals. While the former may be the more desirable for even and efficient pollination it sometimes involves practical difficulties and more labour in transporting and attending to hives. Bees should never be located more than a quarter of a mile from the trees they are required to pollinate for bees do not fly long distances from choice, especially early in the season.

The amount of cross pollination in fruit crops through wind has been proved on many occasions to be negligible, the pollen of all fruit blossoms being unsuited for wind pollination. This applies not only to tree fruits (with the exception of the mulberry which is wind pollinated) but to bush and berry fruits also. The pollen grains of currants and gooseberries are notably sticky and adhere together in masses, making wind pollination quite impossible.

There is a belief that fruit blossoms that have been pollinated are better able to withstand frost, should a sudden frost occur, than those that have not yet been pollinated.

Although fruit crops may be small and unprofitable without

proper pollination there are many other factors that also have a bearing on fruit yield, such as the health and vigour of the tree, pests and diseases, rainfall and climatic conditions generally. Without proper pollination fruit will not set, without proper nutrition it will not grow. There are some abnormal or freak fruits that require no pollination at all, such as the navel orange, the seedless currant grapes and some persimmons, but these are all products of warmer lands.

In Britain the role of the honey bee in pollination is probably more important with tree fruits than with any other class of crop. This is largely because the tree fruits flower early in the season, when the total number of pollinating insects available is very much lower than it is later in the year.

Farm crops are probably of next important after fruit in the value of work done by the hive bee. Large quantities of clover and sainfoin are required for pastures and fodder every year, both crops being largely pollinated by honey bees. In the case of Alsike clover, the seed of which has been mainly imported from Canada in the past, and also in the case of white clover, the increased yields of seed resulting from hives in close proximity to fields has been repeatedly demonstrated. In some cases yields of 2 and even 3 bushels of seed have been obtained by artificially increasing the honey bee population, where formerly only one bushel of seed per acre was obtained.

The hive been may also be an important pollinator of red clover, especially where bumble bees—the natural pollinators of the plant—are lacking. It usually obtains only pollen and no nectar from the flowers and may visit 200 to 300 flowers for a pollen load. In some European countries seed yields have been almost trebled by the use of hive bees, 3 to 5 hives per acre being recommended by one investigator under Danish conditions. In such cases it is necessary that other nectar plants, particularly white clover, should not be available in quantity in the neighbourhood to act as a counter attraction. It has been found with red clover in Australia and elsewhere that maximum pollen collection, and therefore maximum pollination and seed yield, by honey bees, can be effected by feeding the bees liberally with sugar syrup, thereby inducing much brood production and a greater need for pollen. The pollen in the hive may also be maintained at a low level by periodically removing frames with pollen in order further to stimulate pollen collection in the field.

Field beans, mustard and buckwheat are other common farm

crops that are freely pollinated by bees, all being, incidentally, useful nectar and pollen sources. The sunflower has attracted attention as a possible oil seed crop in Britain and has already been grown on a field scale. In Russia, where it is cultivated on a vast scale, the honey bee is regarded as an invaluable pollinating agent and conducive to high yields.

Bees are equally useful to the professional seed grower, particularly the large-scale grower of vegetable seeds. The same applies to the producer of flower seeds.

In the vegetable garden or market garden and in the glasshouse the honey bee may also fill a useful role, particularly in the case of the gourd plants such as marrows, pumpkins and cucumbers, where separate male and female flowers exist and adequate pollination is essential before fruit can develop. Even in the case of indoor tomatoes useful results have been obtained with bees, the flowers yielding pollen but no nectar.

Throughout the country as a whole there is a very uneven distribution of the bee population. In many large towns and in the suburbs of big cities there are often many small beekeepers and overstocking undoubtedly does take place. On the other hand, in many country districts few hives are kept in spite of a fair abundance of clover and other useful nectar plants. It is not uncommon to see pastures well stocked with white clover or bird's-foot trefoil in flower and with few or no hive bees in attendance, even in bright, sunny weather.

ARTIFICIAL BEE PASTURAGE OR PLANTING FOR BEES

The question of planting solely for bees or of providing artificial bee pasturage in order to increase the honey crop is one that frequently occupies the minds of beekeepers. Unfortunately, it is not, or has not proved to be so far, an economical proposition to cultivate any plant on a field scale solely for bees, the cost of tillage, seed, weeding, etc., and the rent or value of the land, far outweighing the value of the increased honey harvest. Where it is possible to make use of the resulting crop in some other way, such as for hay or for seed, the increased monetary returns may make the proposition worth while in the case of farmer beekeepers, assuming of course that there are not large numbers of bees belonging to other people in the neighbourhood.

Among the farm or field crops grown in the British Isles, and known to be good nectar sources, are all the clovers (with the possible exception of red clover), sainfoin, field beans, mustard, rape and buckwheat. There are one or two other crops grown on a field scale, but normally of very local distribution, that afford good bee fodder, such as chicory, teasel, lavender and some of the culinary and medicinal herbs. Unfortunately, with many of the clovers, cutting must take place before flowering, or before flowering is completed, if the best hay or best agricultural use is to be had from them, a good example being crimson clover. Buckwheat is a useful and easily grown crop for light soils, even the poorest, and may be sown at intervals to provide a long nectar flow where sufficient land is available. The seed is always marketable and has many uses, especially for poultry. Unfortunately, the honey from buckwheat is very dark and not of the best quality, although well liked by many people (58).

Planting on a field scale for bees must obviously remain within the province of the farmer or farmer beekeeper, but the establishing of useful nectar plants in waste places falls more within the scope of the small-scale beekeeper or enthusiasts of the hobbyist class, particularly when they are in rural areas and the outskirts of towns and villages. Often disused gravel pits, quarries, rubbish heaps, dumps of various kinds, the sides of newly constructed roads, mounds of soil cleared from drains and ditches, etc., which normally become covered with a useless weed type of vegetation if left to themselves, are capable of supporting useful nectar plants, provided they are taken in hand at the right time. In war-time this may even apply to disused gun sites and temporary camps. Few plants can equal *Melilotus* or sweet clover for establishing as bee fodder in such places. A few seeds sprinkled on any loose soil surface will grow and once established the plant comes up freely from seed. In fact it grows like a weed and is regarded as such in some countries (see Melilotus). Its long flowering season is only one of its many valuable characteristics. Ordinary white clover (*Trifolium repens*), with its creeping and spreading habit, is also very well adapted for naturalizing, especially on gravelly soil. Viper's bugloss has been used in Europe, notably along some of the newer motor roads, for improving the nectar flow. Other easily grown nectar plants suitable for naturalizing in most localities and situations, and usually able to look after themselves with little or no attention, are catnip, motherwort, white horehound, loosestrife, willow-herb, and possibly borage. In shallow or stony soils where few other plants succeed the common blackberry or bramble will generally thrive and prove useful as a late season minor source of nectar and pollen. As a more or less all-the-year-round source of pollen if required gorse may be equally useful in such situations.

A more permanent type of artificial bee forage is available in the form of trees and shrubs that are useful nectar sources as soon as they reach the flowering stage (56). The snowberry (*Symphoricarpos*) is probably one of the best of these for it grows well and quickly in almost any soil. It may not produce flowers as abundantly as do some bee plants but this is compensated by its long flowering season and the fact that the individual flowers are very rich in nectar. Buckthorn has been recommended on the Continent for establishing along roadsides to improve bee pasturage. Some of the cotoneasters might also be well suited for the purpose. The false acacia or robinia is willing to grow almost anywhere, but unfortunately it is too erratic as a nectar yielder in the climate of Britain. Willows are well suited for naturalizing, especially along streams. They may be quickly established from large cuttings and are a valuable early source of pollen and nectar. The sycamore, which is easily established and quick growing, is well suited for waste areas, the margins of fields and pastures, etc., where there is room for it.

The establishment of relatively small patches of good nectar-yielding plants in waste places may not of course result in an appreciable increase in the honey crop, and flowering may not coincide with the main honey flow of the district. Nevertheless, such plants can be of great value in assisting bees to tide over periods when little nectar is available from other sources and thereby reduce the drain on the hive's stores. Furthermore, it must be remembered that every drop of nectar brought into the hive, from whatever source, will assist in producing future foragers that will in their turn bring many more drops to the hive.

The tree planting that continually takes place in parks and open spaces to replace old and decayed trees or in developing new districts and housing estates could be made to improve permanently the neighbourhood for beekeeping if some regard were paid to the nectar value of the different kinds of trees available for planting. Perhaps in the future this may be done, although little if any regard has been paid to it in the past. Nectar value means of course food value and the increased attention now paid to home-produced food may mean a better realization of the value to the community at large of trees that are good nectar yielders. So far as the nectar-producing trees available for cultivation in Great Britain are concerned the limes undoubtly head the list by a long way. Their merits and demerits are fully discussed elsewhere (see Lime). Other useful nectar yielders among ornamental trees are the almond, cherry plum, all the tree fruits, single-flowered cherries and crabs, maples, horse-chestnuts, *Crataegus* and *Catalpa*. These are dealt with under their respective headings in Section 3.

The advisability of planting nectar-yielding trees along the newer arterial roads of Britain that naturally carry very fast motor traffic has been criticized (*Bee World*, 1938, 122), and probably rightly so, on the grounds that it is likely to result in heavy loss of bee life through contact with vehicles. What motorist has not noticed in the summer how dead insects collect in the front of the radiator and how the windscreen is sometimes sticky from the nectar-filled bodies of bees or other insects. Another argument is the risk of increasing accidents by attracting large numbers of bees on to the major highways. It is an easy matter for a bee to get into the muffler of a fast moving motor cyclist or into a saloon car, and be responsible for a sting at an unfortunate moment.

Opinions may differ regarding the force of these arguments, but all will agree it is a pity not to make use of the miles and miles of existing and future arterial roads for improving the available bee forage, especially as this could be done in complete accordance with the general policy of road beautification. On those arterial roads where the trees could be separated from the highway itself by cycle tracks and sidewalks or the trees placed an equivalent distance from the carriage-way, this may be found sufficient to keep most of the foraging bees away from the traffic. On minor roads the objections would hardly apply.

In many parts of Britain there is a dearth of nectar-yielding flowers in the first two or three weeks of June. Fruit blossom (apple, cherry, pear and plum) is a valuable source of early nectar in most areas, tree fruits being so extensively grown in orchards and gardens up and down the country. The wild cherries might be included with them. When these cease flowering other common trees may take their place as nectar yielders, such as sycamore, holly, horse-chestnut and hawthorn. When these have finished flowering, which is about the

last week in May in many districts, there is little else available for
bees until white clover or limes are out. Clover may be in flower by
mid-June or the third week of the month, but produces little nectar
at first. It is usually the last week in June or early July before the
blossoms of the common lime are fully out.

This dearth period, or June gap as it has been termed, in the nectar
flow, is liable to be harmful to the beekeeper. The break in the honey
flow causes the bees, busy with brood rearing at this time, to draw
heavily on the stores procured earlier in the season. This means of
course a reduced final honey yield for the beekeeper himself. The
break may also be an incentive to swarming if it is pronounced. The
presence of minor nectar sources from wild or garden plants or
certain weeds may alter the severity of the dearth in some districts.

Two farm crops that help to bridge this gap and which are usually
in flower in early June, are sainfoin and crimson clover or trifolium
(see Clover). Both are good nectar sources. Unfortunately, neither of
these forage plants is extensively grown. Sainfoin is mainly limited to
chalk districts, while crimon clover, which is an annual and usually
autumn sown, is only hardy in the southern part of the country.

Various garden plants, some of them good nectar yielders, may be
in flower at this time, although such plants are unlikely to be present
in quantity in any district (57). Several annuals favoured by bees
may be in flower at about this period if sown at the appropriate time.
Among perennials, the common catmint is in full flower at this
period and is always amazingly popular with bees for nectar. The
plant has the advantage of being easy to grow and to propagate and
thrives in almost any soil. It also has a long flowering season. Any
beekeeper wishing or able to plant against the June gap might well
consider the merits of this plant, and anyone able to put down an
acre or two as bee forage should be in a position to make interesting
observations, particularly as to the nature of the honey and the
length and intensity of the flow, etc.

Among decorative trees and shrubs that are in flower in the Kew
area during this early June dearth period, and which are visited by
hive bees for nectar, the writer has observed the following (56, 58):

Trees

   Aesculus indica, Indian horse-chestnut (India).
   Cornus alba, dogwood or cornel (N. Asia).
   Cornus occidentalis, dogwood or cornel (W. North America).

Frangula alnus, alder buckthorn (native).
Ptelea trifoliata, hop tree (Canada).
Rhamnus cathartica, buckthorn (native).
Rhamnus purshiana, cascara (W. North America).
Sorbus intermedia, Swedish whitebeam (Europe).

Shrubs

Berberis spp., barberries (several).
Buddleia globosa, yellow buddleia (Peru).
Cotoneaster horizontalis, cotoneaster (China).
Cotoneaster microphyllus, cotoneaster (Himalaya).
Cotoneaster spp., cotoneaster (various).
Erica mediterranea, tree heath (Spain).
Escallonia 'Langleyensis', escallonia (garden hybrid).
Lonicera spp., shrubby honeysuckles.
Pyracantha coccinea, pyracantha (Asia Minor).
Symphoricarpos albus, snowberry (North America).

GARDEN FLOWERS AND THE HONEY BEE

Many of the most showy garden flowers are of no use whatsoever to the honey bee, providing no nectar and little or no pollen. This applies to some of the best roses, chrysanthemums and dahlias, etc., and to numerous other plants that have been 'made double' by cultivation. It is thus possible to visualize a district gay with flowers, such as one where certain cut flowers are grown on a large scale, but one in which the honey bee would starve. Fortunately, nature does not work this way and in most areas there are sufficient nectariferous plants to enable the honey bee to earn a living if not to yield surplus honey for mankind (57).

Another large group of garden plants which are useless to the honey bee are those in which the nectar is so deep seated as to be out of reach, although available to long-tongued insects like bumble bees and butterflies. Sometimes such plants may be useful for pollen or if bumble bees have punctured the bases of the flowers some nectar may be obtainable by this means. It is surprising how often this occurs.

On the other hand, there are many garden subjects which are known to be first-class bee plants. Although these can never compare with wild or crop plants in regard to numbers, they may prove very useful to the beekeeper, especially when they flower at a time when nectar or pollen from other sources is scarce.

Generally speaking, the most useful garden plants are those that flower very early in the year, supplying nectar or pollen at a time when they are very scarce both inside and outside the hive, or else those that flower in the autumn. These furnish food for winter use.

Annuals

A number of garden annuals are useful nectar and pollen plants. They are for the most part easily raised and provide an abundance of bloom while they last. The flowering period can often be much extended by successional sowing and early flowering obtained in the case of the more hardy kinds by autumn sowing. No collection of bee plants should be without a selection of them.

Most garden annuals are not happy unless sown in full sunshine. The beekeeper needs to remember also that plants in shade are often ignored by bees while the same kind of plant a few yards away but in sunshine may be freely visited. The following annuals are among those that are attractive to honey bees, in a few cases for pollen only.

Alyssum, balsam (*Impatiens glandulifera*), calendula, candytuft, China aster, clarkia, collinsia, convolvulus, coreopsis, cornflower (*Centaurea cyanus*), *Cosmos*, *Eschscholtzia*, French marigold, gaillardia, gilia, godetia, gypsophila, lavatera, limnanthes, *Linum*, mallow, *Malope*, mignonette, nasturtium, nemophila, nigella, *Perezia multiflora*, phacelia, pheasant's eye (*Adonis*), mainly pollen, poppy, various kinds for pollen only, saponaria, scabious, senecio, *Statice*, sweet sultan, sunflower, *Viscaria*, *Zinnia*.

Perennials

Many garden perennials are good bee plants. As a group perennials take first place in general popularity with garden lovers on account of the little attention they require when once established.

The following are some of those that are always attractive to bees. They are dealt with individually under various headings in Sections 2 and 3. Achillea, alyssum, anchusa, arabis, aubrieta, campanula, Canterbury bells, catmint, centaurea, cynoglossom, *Epilobium*, erigeron, eupatorium, fuchsia, *Geranium*, geum, golden rod, gypsophila, *Hedysarum*, helenium, hollyhock, horehound, hyssop, lavatera, lavender, *Lythrum*, mallow, marjoram, Michaelmas daisy, *Myosotis*, peony (single), *Rudbeckia, Salvia x superba,* scabious, sidalcea, *Statice,* thrift, thyme, veronica.

Bulbs

Among the early spring flowers so useful in supplying fresh pollen and perhaps a little nectar to the hive after the long winter months, certain bulbous and tuberous plants come well to the fore. Several occur wild in fields and meadows, such as the snowdrop, snowflake and daffodil, while many others are grown in gardens.

The hardy bulbous plants whose flowers are known to be visited by bees include the following: *Camassia* (quamash), *C. scilloides* and *C. cusuckii* especially—mainly summer flowering; *Chionodoxa* (glory of the snow); *Colchicum* (autumn 'crocus')—flowers in the autumn; *Crocus; Eranthis* (winter aconite); *Fritillaria*—including *F. imperialis* (crown lily) and *F. meleagris* (snake's head); *Galanthus* (snowdrop); *Hyacinthus* (garden hyacinth); *Ixia,* in mild areas; *Leucojum* (snowflake); *Muscari* (grape hyacinth); *Narcissus* (daffodil and jonquil); *Scilla; Trillium* (wood lily); *Tulipa* (tulip occasionally for pollen).

None of the above are likely to be sufficiently abundant to affect the honey yield. However, many are easily naturalized, will grow well in shrubberies, under trees, or on the edges of lawns, requiring little attention from one year to another and not competing with other garden plants, but affording their owner an opportunity of watching the bees at work.

Flowering Shrubs

A number of everyday flowering and decorative shrubs are useful sources of nectar to the hive bee. Examples are the cotoneasters, snowberry (*Symphoricarpos*), yellow buddleia (*B. globosa*), pyracantha, escallonia, privet, buckthorn, barberries, elsholtzia, shrubby honeysuckles, flowering currants, heaths, perovskia, callicarpa, aralia, tamarisk and fuchsia, the last mentioned being suited to the milder parts of the country only.

The Vegetable Garden

The flowers of a number of everyday vegetables are attractive to bees for nectar and pollen. Some of these vegetables are normally harvested before they reach the flowering stage, such as radishes, turnips, swedes, cabbage, cauliflower, kale and other brassicas, onions, leeks, chives, carrots, parsnips, chicory and endive. Quite often, however, they are to be seen in flower, through neglect or lack of time on the part of the owner, or if grown for seed. Where seed production on a large scale takes place they may be of appreciable honey value. Other vegetables, which flower during their normal life span in the vegetable garden, and which attract bees, are asparagus, marrows, pumpkins, cucumbers or other gourds and broad beans. The tassels of sweet corn are worked for pollen in some countries but are of doubtful value in Britain (58).

Herbs

Many culinary and medicinal herbs are good nectar plants and afford useful bee forage when grown in large plots in market gardens or on a field scale and allowed to reach the flowering stage. Unfortunately for the beekeeper they are often cut or harvested before flowering takes place—as is the case with mint and sage.

Nearly all the herbs of value to the beekeeper for nectar are members of the mint family (*Labiatae*) and are of an aromatic nature. These include: mint, sage, thyme, marjoram, hyssop, savory and basil. Lavender, rosemary, horehound and catnip might also be added (see Section 3).

BEE GARDENS

A bee garden or plot devoted entirely to plants that are attractive to bees can be the source of much pleasure to any beekeeper interested in plants and gardening, who has the necessary ground or garden space at his disposal (57). The general lay-out of such a garden, and whether it is of the formal or informal type, depends of course upon the inclinations of the beekeeper and the size and surroundings of the ground available. Two important points to be kept in mind are that such a garden should be in full sunlight and not shaded by trees or buildings and that the plants should be grown in bold groups or patches. If plants are grown singly or only in twos and threes they often fail to attract bees even though they be species known to be attractive to them (56). Flowers in shade are often neglected by bees, while those of the same kind of plant in sunlight may be freely worked. Apart from this, the majority of bee plants grow best in full sun in the climate of Britain. It is also advisable not to place plants in very close proximity to hives for often the inmates of these hives ignore such plants. This may be because the bees realize by instinct that plants in the immediate neighbourhood of their hives are liable to be contaminated from their own cleansing flights. In just the same way they generally refuse to take water from troughs placed near the hives.

Many good bee plants are highly decorative and some are suitable as cut flowers for the house—providing always of course that cutting is done with discretion. By a judicious selection of subjects it is possible to have bee plants in flower from early spring until the frosts arrive. A bee garden also enables one to observe bee behaviour among plants of entirely different flower structure.

Most bee plants are fairly accommodating as to soil and will do well in any average garden soil. There are a few, however, that require special consideration in this respect. The heaths and most of the heath family (*Ericaceae*) require lime-free soils with an abundance of peaty or organic matter. On the other hand sainfoin needs an abundance of lime. Most of the clovers in fact and many labiates, such as thyme and lavender, thrive best with a fair amount of lime present. Beds with specially treated or made up soil can often be arranged for such plants.

Among the early spring-flowering plants that are so eagerly sought for pollen, crocuses and the winter aconite should always be included. Snowdrops, Siberian squill and other bulbs are also useful. Annuals offer a very wide choice, some of the best for a bee garden being mignonette, sweet alyssum, phacelia, limnanthes, borage and buckwheat. An equally large range exists among perennials, one of the best garden perennials being *Salvia x superba*, a bed of which will flower for a long time and swarm with bees whenever weather is suitable. Where an edging is required the common catmint has few equals and also has a long flowering period. Chives are also useful and may be used for culinary purposes as well.

A few shrubs are desirable in a bee garden, where there is sufficient room, and require little attention once established. They may often be well placed as a background. Hazels and shrubby willows are useful for early spring where space allows, so also are some of the shrubby honeysuckles, especially *Lonicera standishii*, worked for nectar and pollen. Leter in the year *Cotoneaster horizontalis* never fails to attract (as do other shrubs of this genus) and may be effectively trained to cover walls or wooden fences. The yellow buddleia (*B. globosa*) from Chile is another early summer favourite and easily grown (58). If the locality is one well suited for heaths these may be selected to give flowers more or less throughout the year. All the single-flowered hardy heaths with short corollas or flower tubes appear to be good bee plants. Among the late-flowering shrubs *Perovskia atriplicifolia*, an exceedingly decorative Afghan shrub bearing masses of mauve flowers, and *Elsholtzia stauntonii* (heather-mint) from China are well worth consideration.

There are of course many other plants available for selection, and whatever the size of the bee garden it should be possible to change some of the plants from time to time in order to add variety and interest (57). Apart from the plants already mentioned various herbs and aromatic plants are well worth consideration, such as lavender, rosemary, mint, pennyroyal, sage, thyme, marjoram, savory, hyssop, basil, catnip and horehound. Other plants are purple loosestrife, figwort, willow-herb, mallow, motherwort, gipsywort and catnip—all wild or native species. Single poppies of all kinds are attractive for pollen. Among the taller-growing plants are melilotus or sweet clover, teasel, echinops, rudbeckia, sunflower and hollyhock.

APIARY HEDGES AND WIND-BREAKS

Wind-breaks or hedges are very desirable for the protection of apiaries, particularly in situations exposed to strong winds. In winter they do much to lessen mortality among bees and to result in stronger stocks in the spring. In summer they enable heavily laden, homecoming bees to reach their hive entrance and alight without difficulty. They also afford greater comfort when manipulating in keeping wind off frames and bees, especially early in the season.

While high brick or stone walls may afford the best protection, they are generally out of the question for obvious reasons and the beekeeper looks to vegetation, the cheapest of all wind-breaks, to fill the breach. Evergreen shrubs and hedges are preferable to those that cast their leaves and likely to be far more effective in the winter months when they are mainly needed (56).

Among the well-known hedge plants holly makes the best ever-green hedge under English conditions but is likely to be ruled out by most beekeepers owing to its very slow growth. In some cases, however, it is possible to plant a quick-growing temporary hedge some feet away from the holly hedge, to be removed when the latter has made sufficient growth. The common cherry laurel makes a very effective evergreen hedge or screen and is of fairly rapid growth and easy to grow in nearly all soils, including poor, acid, sandy soils. As a hedge it is best grown with a good wide base and can be tapered towards the top. This prevents the base from becoming scraggy and bare of leaves. It should never be clipped for this mutilates the large leaves but the excess growth cut out with secateurs or a knife. As an informal screen and left to itself it is likely to remain effective for many years. However, where an informal or unpruned screen is desired and there is ample space the common pontic rhododendron (*R. ponticum*) is hard to beat. Furthermore, plants are easily obtained and cheap, if they have to be purchased, but they are often available in quantity in woods. Its rate of growth is much less than that of the laurel, being about a foot a year, and it does not succeed on such a wide range of soils, being averse to heavy clay. Its spreading branches maintain a good cover right down to the ground. This is what the beekeeper wants for his hives are at ground level. Laurustinus (*Viburnum tinus*) is another evergreen of somewhat similar habit and at the same time useful for early pollen, but again of very slow growth.

When a hedge is required primarily as a barrier to keep out intruders or livestock few plants can equal the common hawthorn or may (often called 'quick'), when properly grown and managed. Although without leaves in winter, close clipping induces a dense mass of woody shoots and this offers fair resistance to wind. Among the many good points that have made it the premier utility hedge of the country for centuries, in spite of the introduction of numerous potential hedge plants from other countries, are its rapid rate of growth, ability to stand close clipping, hardiness and suitability for most soils. The fact that it is easily raised from seed, or is cheap to buy as seedlings and transplants well, are other virtues.

Several evergreen barberries make first-class hedges and are also effective barriers, the best known being *Berberis x stenophylla* and *B. darwinii*. Regular pruning each year with these plants, immediately after flowering, helps to prevent the base from becoming bare. Pyracantha may also be used where a prickly hedge is required, both the common sort (*Pyracantha coccinea* 'Lalandei') and *P. rogersiana* from Yunnan, being suitable. The latter quickly makes a hedge six feet high. Pyracantha has the drawback of transplanting badly, even when quite young. Another prickly but perhaps rather indifferent hedge, although well suited for poor sandy soils of the heathland type, is gorse. It withstands severe winds remarkably well and may be close clipped. Seed is generally sown direct for it transplants badly. It will be noted that all these barrier plants happen to bear flowers that are of nectar or pollen value to the hive bee.

Other evergreen hedge plants whose flowers are visited by the honey bee for nectar are cotoneaster, escallonia and tamarisk. The last mentioned is invaluable in exposed situations near the sea and withstands salt-laden winds as few other plants will do. It may either be grown naturally as a shelter belt or trimmed as a hedge. In the latter case it should be clipped in the spring so as not to interfere with flowering. Most cotoneasters are first-class nectar plants and some make good hedges. The best known is *Cotoneaster simonsii*, which may be grown as a hedge up to 6 feet and is well suited where a hedge without great depth is required, as in small gardens. On the other hand, where there is space for ample depth in the hedge, to add to effectiveness as a screen, some of the escallonias are well adapted, particularly in the milder parts of the country as in the south and west coasts. There they are much favoured with their dark, glossy leaves and free-flowering habit. One of the best as well as one of the hardiest is *E*. 'Langleyensis' (see Escallonia).

The virtues and drawbacks of privet as a hedge are familiar to most people. Its quick growth may commend it to some and it is usually semi-evergreen but most beekeepers are too busy for the constant clipping required to keep it in good shape. If left to its own devices it soon assumes tree form, becoming thin at the base and flowering freely every year. The flowers are a good source of nectar but honey from a privet has a bad reputation and may spoil other honey.

The common beech makes a good hedge and wind-break and succeeds on chalk soils. It stands close trimming while the retention of the dead leaves in winter adds to its attraction and shelter value. The purple beech may be used in the same way.

Yew also grows well on chalk, but like holly and box, is probably too slow growing to interest the beekeeper and is prone to attack by scale insects. Box has the advantage of succeeding in partial shade but is liable to become bare at the base.

Some conifers are useful as quick-growing wind-breaks but generally become open at the base after some years. Lawson's cypress (*Chamaecyparis lawsoniana*, formerly *Cupressus lawsoniana*) and its many varieties is perhaps the most generally useful. The

Monterey cypress (*C. macrocarpa*) makes a good informal screen in the milder maritime districts. Arbor-vitae (*Thuja occidentalis*) thrives on good soils but is not happy on poor land.

Frequently the beekeeper requires a quick-growing screen that will be effective very early or in the same year in which it is planted. This may apply in the case of temporary out-apiaries or where hives are situated near a footpath or road and it is desired to divert the line of flight of the bees to a height of several feet soon after leaving the hive so as to avoid the possibility of collision with passers-by! The Jerusalem artichoke (*Helianthus tuberosus*) is sometimes used with good effect in this way. Its rapid rate of growth enables it to attain a height of several feet quite early in the season and no plant could be less exacting in its requirements. It also has its uses in providing summer shade for hives that are completely exposed to the sun, particularly hives of the single wall or 'National' type. A few tubers planted on the south or south-west side of a hive result in effective midday and afternoon shade during the hotter months of the year. Another quick-growing summer screen is *Reynoutria sachalinensis* (formerly *Polygonum sachalinense*), which sends up its tall leafy stems year after year when once established. The climbing polygonum (*P. baldschuanicum*) is also useful given the necessary support. The more vigorous cultivated blackberries, such as 'Himalaya Giant' and 'John Innes' may be used in similar fashion, but should not be too near hives for the thorns on their spreading branches will only too readily tear open a bee veil, perhaps at an awkward moment.

The following suggestion for an apiary hedge which appeared in a Dutch beekeeping journal (see *Bee World*, 1931, 65) may commend itself to some beekeepers. 'In spring plant out strong willow cuttings one yard apart. If the ground be moist enough they will grow. In autumn tie them together so as to form a flat hedge. Next spring between every two willows plant a bramble, preferably a large fruiting kind, and wind and tie the stems along the hedge. In a few years one has a splendid wind-screen, needing no attention save pruning. Birds nest in it freely preferring it to nest boxes and it is a joy to the bees in both spring and summer. March is the best season for planting willows' (see Willow) (47).

HONEYDEW AND PROPOLIS

As honeydew and propolis are only too familiar to all beekeepers and are derived from plants a few words in regard to them and their botanical sources may not be out of place (39, 55).

Honeydew is common in some seasons, particularly during hot dry summers, and will spoil the quality of honey if present in any quantity. Various insects feed by sucking the sap or juices of plants and excrete a sugary liquid which is little other than a modified form of plant sap. Bees will collect and store this fluid or honeydew, but usually only if ordinary or floral nectar is scarce. The insects responsible include plant lice or aphids, scale insects and leaf hoppers. They are sometimes very numerous and breed with great rapidity. When exuded the liquid is colourless and sweet and quite whole-some, any objection to it being quite ill founded. On drying, however, it becomes sticky, being of a gummy nature, and usually infected with fungi or 'sooty moulds' giving it a black or dark coloured appearance. It is these moulds that darken the honeydrew or honey with which it becomes mixed.

Most of the common trees of Britain may harbour insects of this kind and so produce honeydew. It is only in certain seasons that it is prevalent. These trees include the lime (the worst offender), oak, sycamore, beech, elm, ash, chestnut, hawthorn, fruit trees and varius conifers. A number of cultivated or garden plants may also be affected.

Where limes are plentiful the honey may be completely spoiled in some seasons, especially section honey. Fortunately this does not often occur. When honeydew has been collected in quantity honey is dark or olive-green in colour and has a taste suggestive of treacle. It does not granulate. This is due to the large percentage of dextrin or gum present, sometimes as much as 10 per cent as compared with less than 1 per cent in most ordinary honey. This high percentage of gum makes it a bad food for wintering bees as the gummy matter is prone to clog the intestine if frequent cleansing flights are not possible. Although most honeydew honeys are considered to be safe for spring stimulation the honeydew from limes is now considered to be inherently harmful to bees.

In hot dry seasons when bees are collecting honeydew, other sources of nectar being scarce or dried up, they are sometimes inclined to be cross. The reason put forward for this (20) is that the honeydew gets too hard and sticky during the heat of the day for the bees to collect but is softened by dew at night. It is thus available in the morning but not later in the day and the sudden cessation accounts for the bad temper.

Some races of bees are more prone to collect honeydew than others and this explains why some hives in an apiary of mixed bees may be found to have much honeydew but not the others.

Honeydew honey is much more prevalent in certain other countries than it is in Britain. In parts of Europe it is obtained in quantity from conifers while in Hawaii much of the honey produced consists of honeydew from insects that feed on sugar cane. Imported West Indian honey often consists largely of honeydew. Such honey finds uses in the confectionary and baking trades.

Among the conifers known to produce honeydew are pines (*Pinus*), firs (*Abies*), spruces (*Picea*), cedars (*Cedrus* and *Libocedrus*), larches (*Larix*) and junipers (*Juniperus*). Generally it is not produced every year but only in those seasons when conditions are favourable for it. The honey obtained from this source is generally thick and dark, even greenish-black, and with a characteristic strong flavour. It is well known in many parts of Germany and may form half the honey crop in some instances, also in Switzerland. Known as 'tannehonig' it is often preferred to ordinary honey by local inhabitants or those accustomed to it. In Germany larch (*Larix decidua*) and silver fir (*Abies alba*) are considered to be the main sources: pine only to a lesser extent. The incense cedar (*Libocedrus decurrens*) is a source of honeydew honey in California.

This conifer honey is known in Britain, but is only rarely obtained. It would seem that the conditions necessary for the honeydew to be produced by the trees rarely accompany an English summer. In parts of Central Europe it is only obtained in hot summers. It has been recorded in the Camberley area of Surrey (by no less an authority than Miss Betts), where Scots pine is very prevalent in the neighbouring heathland (Bagshot Heath). Comb honey from it was described as uncommonly good, but with cappings dark and watery looking, and not attractive (*Bee World*, 1936, 7). It has also been recorded from Scotland (Ross, Inverness and Aberdeenshire), from the same botanical source, the honey being stated to be favoured by connoisseurs of honey and very dense, although easy to extract (*Scottish Beekeeper*, August, 1938).

This type of honey, although apparently superior to most honeydew honeys for eating purposes, is considered to be bad for wintering. It probably has a high protein content like other honeys of this class.

Propolis

Propolis or bee glue as it is sometimes called is a nuisance to every beekeeper, particularly towards the end of the season when it is most in evidence. Its primary use by bees is to stop up cracks and crevices in the hive. It may be collected from a number of different plants. Some of these are known with certainty but there is doubt regarding others. Bees have been observed tearing away lumps of the resinous matter from plants, especially the trunks of pine trees, with their mandibles and eventually packing it in their pollen baskets. There is still much to be learned, however, regarding the plant sources of propolis in the British Isles throughout the summer months. Unfortunately the collection of propolis by the honey bee is difficult to observe and is not the spectacular process that the collection of pollen is.

Propolis varies a good deal in colour, also in physical properties, although the more obvious qualities such as stickiness and brittleness are largely governed by temperature. The extent to which it is used is also largely dependent upon the variety or race of bee. With regard to colour this may range from yellow to dark reddish-brown according to the plant source. When fresh it may be clear and colourless. The greenish colour so common with propolis in the hive is due to admixture with wax or other substances. Wax is commonly mixed with propolis by bees.

Many plants secrete sticky substances in some form or other. Besides the drops of resin on the trunks of pine trees, familiar examples are the sticky buds of the horse-chestnut in early spring. Pine resin is definitely collected by bees as propolis but in the case of the horse-chestnut the gummy matter is present at a time when bees have no use for it, for in Britain propolis is never collected between the months of October and April. Nor is it collected during a strong honey flow. Some poplars, especially the balsam poplars, produce a resinous material on the buds which bees are known to use as propolis both in this country and elsewhere. Bees have also been observed collecting the gummy matter from the heads of sunflowers and the buds of hollyhocks. Other plants that have been referred to as sources of propolis in the British Isles are: birch, alder, beech, willow, and chestnut.

Bees will of course collect other materials for propolis besides exudations from plants. Old quilts and hives with propolis already attached to them are always welcome on hot days. Bees have also been observed taking bitumen and varnish from woodwork. It is thought the old saying that 'bees follow their master to the grave' is due to bees having been observed on coffins, the fresh varnish being the attraction.

Propolis is not always easily distinguished from pollen on the bees' legs. It is generally present in small loads only. In hot weather they may appear as two glistening dots.

A bee may take from a quarter to one hour to collect a load, and on arrival at the hive other bees remove the loads with their jaws, which may be a laborious process. Sometimes this is done on the floor of the hive or the alighting board, hence the specks of propolis that frequently collect there, some being dropped and left.

Mankind has found few if any regular uses for propolis. It has been used in medicine and in leather polishes. The old Italian violin varnish as used by Stradivarius and other noted makers of Cremona is now believed to have been nothing other than propolis derived from poplar (*Bee World*, 1936, 56). In the wild state bees use propolis to make their hives or nests water- and wind-proof and to restrict entrances to exclude natural enemies but under domestication its use is superfluous as beekeepers will wholeheartedly agree.

Section 2

The Major Honey Plants

CLOVER
Trifolium spp.: Leguminosae

The clovers are the most important honey-producing plants in Britain and are considered to account for about 75 per cent of the yearly honey crop (35). White clover (*T. repens*) is by far the main producer among the different kinds of clover that are cultivated or occur in pastures. Besides the cultivated clovers which may almost be numbered on the fingers of one hand, there are a number of wild clovers which are never cultivated, but which are useful nectar plants. About two dozen species of *Trifolium* are included in the British flora, but some are rare or very local in distribution and so are of little consequence to beekeepers. Several of the cultivated kinds that are good nectar plants occur freely in both the wild and the cultivated state. They are dealt with separately in the discussion that follows. Some of the clover-like plants that afford good bee pasturage, but which are not true clovers and do not belong to the genus *Trifolium*, are dealt with under their respective headings—sainfoin, melilotus or sweet clover, lucerne, bird's foot trefoil, etc. (40).

White Clover   *Trifolium repens*

This valuable bee plant is the premier honey producer in many other countries besides Britain. Large quantities of honey are obtained from it in Canada, the northern United States, parts of Europe, New Zealand, Australia, Tasmania, etc. It has been stated that a greater quantity of honey is obtained from this plant throughout the world than from any other individual plant, also that wherever the Anglo-Saxon race has settled or become the dominant race the mainstay of theiry honey production will be found to be this plant (41). The best grades of imported Commonwealth honey, i.e. Canadian and New Zealand, are derived from it.

Throughout Britain white clover is one of the commonest plants and is to be found in all types of pasture excepting those on acid soils.

It is prevalent along roadsides where it often thrives in the gravel or sand swept off the road surface. This may be due largely to the good soil aeration such a habitat provides. It also occurs freely in waste places and as a weed in arable land and lawns. It even grows well on the slag heaps of some of the midland industrial towns (Black Country) where enterprising beekeepers have wrested good-quality honey from it in spite of the otherwise uncongenial conditions (9).

Being a valuable fodder for stock it is much grown for pasturage, usually in company with grasses and other clovers. It enters into the composition of seed mixtures used for both temporary leys and long duration grassland. There are two main types of white clover—the so-called wild white and the ordinary white or Dutch clover, much of the seed being imported from Holland. The wild white is a smaller plant than the Dutch, with smaller leaves and flowers and a free running or creeping habit. It is also longer lived and is most used for permanent pasture. The seed is more expensive than the Dutch and is collected from finest old sheep pastures in Kent. There is probably little difference between the two types as honey producers, but some observers (13) have expressed the view that the wild white is more freely worked than the Dutch when they are grown side by side. There are many improved strains or varieties of white clover. One which has come well to the fore in English ley farming is S.100 (Aberystwyth) which quickly establishes itself, commences growth early and continues growing late into the autumn. Ladino clover is a giant form of white clover that originated in Northern Italy and has interested beekeepers in the United States (*Gleanings in Bee Culture*, 1943, 700).

White clover commences to flower early in June in most southern districts and continues in flower for the greater part of the summer, provided prolonged drought does not intervene. The actual honey flow usually commences about ten days after the first open blossoms appear. Flowers that appear towards the end of the flowering period do not produce much nectar.

Each flower-head contains from fifty to one hundred individual flowers or florets. These all stand erect at first, but as they become pollinated and cease to secrete nectar they bend downwards and eventually wither. The flower-tube is relatively short and other insects besides honey bees are able to get at the nectar. The stamens or male elements of the flowers are hidden from view and are united into a tube for the greater part, as in other clovers and most leguminous plants. It is on the lower inner surface of this staminal tube that the nectar is to be found.

Under favourable conditions the flowers secrete nectar very freely and heavy honey crops are taken in good seasons. The temperature range over which white clover will secrete nectar is wide compared with that of many plants, and probably extends lower than the temperature required for bees to fly. Assuming there is adequate soil moisture and the plants do not wilt, it is doubtful whether the mild climate of Britain ever becomes too hot for the plant to secrete. Most beekeepers in clover areas hold the view that the warmer the weather the better the flow, assuming always that drought does not unduly deplete soil moisture. This may not, and probably does not apply to Continental climates where much hotter summers are the rule. In the United States it is said of the plant: 'It rarely may be counted upon as a major honey-source where the average summer temperature exceeds 75° F. A more important consideration, however, is that secretion is most rapid where there is a considerable daily range of temperature, the best results being observed when the night temperature is below 65° F. and the daily temperature above that point.' (*Beekeeping in the Clover Region*, U.S. Dept. Agric., Farmers' Bull. No. 1215.)

Even with the right climatic and temperature conditions the soil is all-important in determining to what extent white clover will prove to be a good honey source. Soils having an abundance of lime always yield the best results and some of the best clover honey districts are those on the chalk, such as the slopes of the Chilterns and Cotswolds. Some contend the plant yields better on hilly slopes than it does on lowlands and will continue to yield later in the year (14). Where there is a marked deficiency of lime in the soil, as in the neighbourhood of Bagshot Heath, white clover may yield no nectar at all, but be visited by bees for pollen.

Some of the sheep farming districts of Leicestershire and Lincolnshire afford first-class bee pasturage, white clover being grown on a large scale and often without other mixtures. A mass of bloom over a wide area is not uncommon in such areas. White clover has the advantage that it is invariably grown to be grazed and not to be cut away completely for fodder or for hay just when it is in flower, as is the case with some clovers. This means that there is always blossom available to the bees throughout the season.

White clover honey is the honey *par excellence*—the honey with which all other honeys are compared. It is light in colour, from water white to pale amber, and very bright. It has good density combined with a delicate flavour and aroma. The flavour has a more or less universal appeal. Unlike some honeys, it is not liked by some people and disliked by others. This accounts for the popularity of the honey as a trade product. It finds favour with the largest possible number of customers. The density and colour of the honey vary to some extent with the season and the soil. When the nectar flow is slow or intermittent the honey is darker than when it is obtained from a fast steady flow. The honey granulates well with a fine smooth grain and is white, almost resembling lard. It does not granulate quickly. This is one of the reasons why it is favoured for comb honey, in which form it can hardly be excelled.

The pollen of white clover is pale yellow in colour and is produced rather sparingly by the flowers. When packed in the bees' pollen baskets it assumes a dull greenish appearance and is not usually brought back to the hive in large loads as is the case with many pollens. It is one of the pollens most commonly found in English honey. The individual pollen grains are somewhat variable in size, shape and colour, and may easily be confused with those of certain other clovers and clover-like plants.

Red Clover   *Trifolium pratense*

Like white clover, this species occurs very freely in the wild and the cultivated state and is of great importance to the agriculturalist. It is, however, not nearly such a good honey plant as it is only under special or exceptional conditions that the honey bee is able to work its larger flowers for nectar. The flowers are red or purplish in colour, but otherwise the flower-heads are very similar to those of white clover. The plant may be distinguished when not in flower by the white 'horse-shoe' mark on the upper surface of the leaf.

There are many varieties of red clover used in agriculture. They are employed in pasture seed mixtures of various kinds and may be sown as a sole crop for cutting. They fall into three groups: (1) early or medium flowering (cow grass or broad red type); (2) late flowering or single cut red clover; and (3) the wild or indigenous type. The early flowering sorts bloom from two to four weeks sooner than the late flowering.

When honey has been obtained from red clover it has usually been from the second crop, i.e. the crop arising after the first crop

has been cut. The reason given for this in the past has been that the
flowers of the second (or third) crop are smaller and with shorter
flower-tubes, which enables honey bees to reach the nectar. This
theory has, however, been discredited as a result of careful measurements made of the flower size of first and second crop red clover. It
has been shown that there is really very little difference in tube
length, the small reduction being insufficient to account for the bees'
ability to reach the nectar. The explanation now put forward is that
it is the forces of capillarity that enable the bee to extract the nectar,
provided that in the first place secretion is sufficiently copious for the
nectar to rise appreciably in the tube. The nectar forms a meniscus in
the tube, aided by the presence of the ovary, so that it stands higher
against the side of the staminal tube. Once this reaches a sufficiently
high level for the honey bee to reach it with the tip of the tongue the
bee is able to extract all the nectar. If, however, the nectar rises a
certain amount but remains just out of reach of the bee she is unable
to obtain any of it, although of course bumble bees may take it. In
other words, if the flower holds sufficient nectar for the honey bee
to reach it she will be able to empty the flower, but will get nothing
more from that flower until such time as the nectar again rises
sufficiently high for her to reach the top of it. From this it is clear that
while the longer tongued bumble bees are able to get nectar from red
clover, however little may be secreted, it is only when secretion is
heavy, i.e. conditions for nectar production very favourable, that the
honey bee is able to work the flowers for nectar (*Bee World*, 1936,
102). In many areas conditions for nectar secretion are quite likely
to be more favourable later in the year when second or third crop
red clover is in flower.

It has now been shown by Dr C. G. Butler (*Annals of Applied
Biology*, 1941, 125) that both first and subsequent crops of red
clover may be worked for nectar by honey bees under favourable
conditions and that they may also visit the flowers for pollen only.
The average length of flower-tube is 9·6 mm. and the average honey
bee is estimated to be able to reach 7·9 mm. Therefore the nectar
must rise 1·7 mm. or more to become available. At Rothamsted
'during the dry summer of 1940 this appeared to happen about once
in every four or five days, usually after a night of heavy dew when the
honey bees were found working this plant for nectar. A heavy dew
leads to the absorption of water by the nectar and consequently to a
rise in height in the corolla tube, which may be sufficient to bring it

within reach of the honey bee. A number (127) of measurements were made of the height to which the nectar in the corolla tube rose, and it was found to vary between approximately 0 and 3·4 mm.' (C. G. Butler.)

A special variety or strain of red clover known as Zofka clover, is claimed to have a short staminal tube, only 6·5 to 7 mm., and to be well suited as a nectar plant to the honey bee. It originated from a Dr. Joseph Zofka in Czechloslovakia (*American Bee Journal*, 1937, 478–80).

As the actual amount of nectar secreted by the red clover flower is greater than that from either of the other two commonly cultivated clovers in Britain (white and Alsike clover) it is unfortunate that the honey bee cannot take full advantage of it. Nevertheless red clover may be a more important minor source of nectar than is generally supposed by beekeepers.

When honey is obtained from red clover it is of the same high quality and has the same general characteristics as that of white clover, but may granulate more quickly.

Alsike Clover   *Trifolium hybridum*

Alsike or Swedish clover derives its name from the village or district of Alsike in Sweden where it originated. It was introduced to Britain over a hundred years ago and is much used in seed mixtures for pastures, especially short leys, most of the seed used in Britain coming from Canada. It is a valuable hay plant and is generally grown with grasses for support, being somewhat prone to lodge. This clover has the advantage of being able to thrive under soil conditions that are too wet and acid or too deficient in lime for white or red clover. It is also more resistant to clover sickness than red clover. Unlike red clover it makes little or no second growth after cutting.

The flower-heads are very like those of white or Dutch clover, but more pinkish. The plant is of erect growth, 1 to 3 feet, and intermediate in general characters between red and white clover. Its flowering period is equally long.

The mechanism of the flower in Alsike is the same as that in white clover and it is of about the same value as a honey plant. In some areas in fact it is considered to yield nectar more freely than white clover (20). The honey obtained from Alsike clover is so similar to that from white clover that it is doubtful whether it is distinguishable at all.

Crimson Clover   *Trifolium incarnatum*

This clover, also called Italian and carnation clover, is an annual not a perennial like most clovers. It is a native of southern Europe and was at one time grown in flower gardens on account of its showy, deep crimson, flower-heads, which are elongated and not globular in shape. It is now generally seen as a farm crop. Fields in flower present a magnificent sight and may be conspicuous miles away. The name trifolium is commonly applied to it in agricultural circles in Britain, where it is grown mainly in the south and south-eastern districts, not being hardy enough for the north. It is usually sown in September to stand through the winter, often as a catch crop after a cereal, when the simplest cultivation only is needed.

Flowering takes place early in the following summer before the other clovers are out, and it is here that the value of trifolium to the beekeeper mainly lies. It is a valuable honey plant and often fills the gap in the nectar flow between the fruit and tree blossoms and white clover. It may be fed off with hurdled sheep or to dairy cows. This is done before the plants are in full flower as the flower-heads become somewhat prickly with age and may cause digestive troubles. It is unfortunate for the beekeeper that this has to be done and means that the bees are unable to benefit from all the flowers.

There are several varieties of trifolium—early, medium, late and a white-flowered late variety. Honey from this crop is light in colour and similar to that of other clovers.

Yellow Suckling Clover   *Trifolium dubium*

This tufted annual bears small yellow flower-heads of about a dozen flowers and is common in grassland. The seed is sometimes incorporated in seed mixtures for temporary leys in soils well-suited to cocksfoot, when abundant aftermath is desired. It is useful on poor light land. In spite of their insignificance the flowers are attractive to bees and yield nectar abundantly. They appear in June and July and doubtless contribute to the honey crop in some areas. The plant is also known as small yellow trefoil.

Hop Clover   *Trifolium campestre* (formerly *T. procumbens*)

Hop clover or hop trefoil is another of the small or insignificant clovers common in pastures and the borders of fields. It bears hop-shaped heads of yellow flowers from June to August, which are also a useful minor source of nectar. This clover may be used in seed mixtures to give good bottom cover. It is sometimes confused with ordinary trefoil or black medick (*Medicago lupulina*).

Honey bees also visit the flowers of strawberry clover (*T. fragiferum*) and harefoot clover (*T. arvense*), both wild plants but not cultivated, the latter often common near the sea.

LIME
Tilia spp.: Tiliaceae

As a honey producer the lime or linden is second only to clover in the British Isles. While the clovers are considered to account for about 75 per cent of the total quantity of honey produced, the lime supplies the greater part of the remainder (29, 54).

For many thousands of British beekeepers who reside in or near towns the lime is their sole source of surplus honey. In the aggregate such beekeepers produce a not inconsiderable amount of honey every year for home or local use. The large-scale or commercial beekeeper who relies on pastures or field crops for his honey crop may however prefer to be without limes within flight range of his bees owing to the possibility of honeydew contaminating or spoiling his main honey crop, for the tree is a bad offender in this respect in some years.

There are some thirty different species of lime or *Tilia* known to science. All are natives of the north temperate zone—Europe, North America and Asia, including China and Japan. They all have many characteristics in common, such as type of leaf and flower structure, the flowers being borne on a conspicuous bract. Furthermore, it is probable that all are fragrant and yield nectar, for this applies to about two dozen species that the writer has been able to observe personally in flower, although it would seem that some secrete nectar more freely than others. Apart from botanic gardens or private collections only three limes are commonly met with in Britain. These are, in order of their prevalence:

Tilia x europaea (formerly *T. vulgaris*) the common lime of streets, parks and private gardens; of European origin and a hybrid. Different individual trees may vary somewhat in their time of flowering, even in the same area. They may reach large dimensions—over 100 feet in height—and bear smooth branches and heart-shaped leaves two to four inches long. Trees may reach a great age and many famous and historical trees and avenues exist.

Tilia platyphyllos, wild in parts of Hereford, Radnor and York-shire; very like the above but with larger leaves, downy shoots and fewer flowers on each bract. It is often planted in place of the above and has roughly the same flowering period. Hybrid forms, or forms intermediate between *T. x europaea* and *T. platyphyllos* are not un-common. Like *T. x europaea* different trees in the same area may vary a good deal in their time of flowering, some being several days earlier or later every year, than others. A certain street tree known to the writer in the Kew district always flowers about a week before any other tree in the neighbourhood. If such a tree were propagated (vegetatively) it is probable this valuable characteristic would be maintained.

Tilia cordata, the small-leaved lime; occurs wild or apparently wild in some parts of the west of England. It is characterized by its smaller leaves and flowers which are not all pendent and therefore probably more liable to have their nectar spoiled by rain. It flowers later than the above two common limes but is not often planted. It is much more important for honey on the mainland of Europe than in Britain, for there extensive woods occur and it is commonly planted as a street tree.

A few of the other species of *Tilia* are sometimes planted as ornamental trees such as *T. tomentosa* (the silver lime), *T. petiolaris* (the weeping silver lime), and occasionally *T. euchlora*, sometimes referred to as the Crimea lime.

The two common limes referred to above (*T. x europaea* and *T. platyphyllos*), which supply practically all the lime honey harvested in the country, generally come into flower in the third week in June in the south-east of England but later farther north and earlier in the west. In early or late seasons they may depart from this by anything up to ten days. The flowering period does not generally last for more than two to three weeks or perhaps a month if cold or wet weather intervenes. The individual flowers open at night and last for about a week, turning to a darker shade of cream or yellow on about the third day. Nectar is secreted during the whole of the time and more copiously when the darker coloured or 'female' stage of the flower is reached. For some unknown reason some of the flower buds never open.

The intense fragrance of the lime in flower is well known and even suggests honey. It may often be detected many yards away from the tree. The flowers are commonly collected and dried in European countries and used for making a kind of tea or tisane. It is only rarely that the lime ripens its seed properly in the climate of Britain, requiring a hot or Continental type of summer to do so. This applies also to most of the other introduced limes.

Nectar secretion in the lime flower is rather unusual in that it takes place on the inner side of the five boat-shaped sepals, and is held in position there by small hairs aided by surface tension. The sepal may become only moist with nectar but when secretion is copious as sometimes happens on sultry mornings, the nectar collects in large drops and is to be seen glistening in the flowers. Shaking a branch then may cause quite a shower of nectar. The secretion of nectar in the five sepals of a single flower is not always uniform, some often containing more than others.

In working a lime flower for nectar, especially if it is only moist, the honey bee often scoops the end of the ligula or tongue round the inside of the sepal, removing all traces of nectar. This may easily be seen from above the flower owing to the transparent nature of the sepal. Sometimes a bee will alight on a flower that obviously has nectar, and yet proceed to another flower before taking any nectar. Is this because the sugar concentration in different flowers varies, as with age, and the bee knows how to select the one with the high sugar content?

There has been much discussion as to what constitutes the best climatic or weather conditions for free nectar secretion in the lime. Some interesting work in this connection has been done by Continental workers (Beutler and Wahl), but it is probable the conclusions arrived at cannot always be applied to Britain where conditions are different and average summer temperatures lower than those of similar European latitudes.

Observant beekeepers in Britain with experience of lime districts generally consider the best conditions for a good lime flow are warm nights and warm sultry days (high atmospheric humidity), with the sky perhaps overcast; but certainly not hot, dry, bright sunny days such as might be the most favourable for many other nectar plants. It must be remembered the lime secretes its nectar mainly, if not entirely, before midday. Bees may often be seen working limes very early in the morning. With hot dry days or drying winds there is a tendency for the nectar to be dried out early in the day and little more is secreted, with the result that the bees find little or nothing to bring in during the afternoon and evening. The open nature of the lime flower is also conducive to this. Temperature is of course of great importance for there is little or no secretion in cold weather. In Europe the optimum temperature for lime nectar secretion has been considered to be 66–70° F (*Bee World*, 1938, 29). The same might apply to Britain.

The lime does not appear to be particular with regard to soil and will thrive and secrete nectar in a variety of soils that may differ from one another considerably in physical texture and in chemical composition. While soil moisture may be important in countries with a naturally dry climate it is doubtful whether it is of much significance in the relatively wet climate of Britain. Furthermore, being a tree and deep rooted, the lime is able to draw moisture from a considerable depth in the soil and is not dependent on the moisture in the surface layers like many honey plants that are of short duration.

In most years and in most districts in Britain a certain amount of honey is always obtained from the lime where it is sufficiently abundant, but really good or 'bumper' years only come round once in every three or four seasons. It is seldom that it fails altogether. When it does it is due to unfavourable weather at blossoming time, such as excessive rain or cold winds. Cold winds are definitely inimical to the nectar flow but fortunately are not usual at the time of year when the lime flowers. All winds, however, are not necessarily harmful for the production of nectar. After rain mild drying winds may have a beneficial effect in raising the sugar concentration of the nectar.

In spite of the pendent nature of the flowers and the nectar being secreted on the undersurface of the sepals, where it would appear to be well protected, heavy rain does cause serious dilution of the nectar. This was well demonstrated at Kew in the summer of 1943 by the writer, in collaboration with Mr L. R. Hayward, in connection with work on sugar content of the nectar of different species of lime growing at Kew. Incidentally, this work indicated that there was probably little difference in the nectar concentration of different species of lime flowering at the same time but that the age of the flower was probably important, old flowers having more copious and a richer nectar than those newly opened. In Germany it has been found that there is little difference in the composition of lime nectar

C

from different parts of the country (Beutler). There seems to be some evidence that flowers at the top of a tree may have a different nectar concentration from those near the ground.

To obtain surplus honey from limes it is of course essential to have a sufficient number of trees for the number of hives or bees in the neighbourhood, a few isolated trees being of little account. In this connection it is interesting to note that according to Beutler and Wahl's figures about forty average-sized lime trees are needed per hive for an increase in hive weight of 4 lb. daily (*Bee World*, 1938, 17). Where there are many trees and the flow is heavy bees may get all the nectar they can deal with from trees close at hand and do not bother to visit those further afield. Probably no other plant or tree secretes such quantities of nectar as the lime when at its best.

Lime honey is well liked by most people, in spite of its distinctive flavour, usually described as 'minty' or 'like peppermint'. It is usually light amber with a greenish tinge, which may be due to traces of honeydew, for there is always a certain amount on the leaves of the lime. Its density, however, is not good for it is always thin. It crystallizes after a few months with a fine smooth grain. Lime honey always contains lime pollen to a greater or less extent. This is of a dull yellow colour, the individual grain being like a flattened sphere in shape with three pores on its edge. It is of average size for pollen—25 to 30 microns (27).

The lime is the worst offender among British trees for honeydew. This is collected most by bees when hot dry weather coincides with blossoming time, for then nectar ceases to be available early in the day and the bees turn to honeydew. While plenty of nectar is available they do not seem to bother about it. As the honeydew itself is always darkened by fungus (sooty moulds) it causes the honey to be dark. This may vary from olive green to almost black and the flavour is always inferior, sickly to some. It has been shown that lime honey-dew may be toxic to some extent to bees (*Bee World*, Vol. 24, 6). The flowers of some species of lime (e.g. *T. petiolaris*, *T. tomentosa*, *T. orbicularis*) may also have a harmful effect on bees, for it is common to find dead or dying bees under the trees. At Kew this is much more noticeable in some seasons than others. Casualties are always much higher among bumble than hive bees, probably because of their larger honey sacks and the greater quantity of nectar they consume. In many seasons no dead hive bees are to be seen under the trees, only bumble bees. It is probable the actual amount of poisoning caused by limes is small having regard to the total bee population and that their merits as honey producers far outweigh their demerits.

The lime has been much planted as an avenue tree throughout the country in the past and many fine avenues still exist, especially as approaches to country mansions. It has also been much used in street planting but there has been a falling off in lime planting in the last few decades for various reasons. This is unfortunate for the beekeeper, or rather for future generations of beekeepers. Present-day beekeepers benefit from the lime planting of their forefathers, the lime being a long-lived tree and requiring ten to fifteen years before it commences to blossom as a rule. Some of the newer limes, however, and those from the Orient do not have the weaknesses of the common lime for general planting although quite hardy and equally good as bee pasturage. If these trees could be given a fair trial as street and avenue trees, and for parks, pleasure grounds and open spaces generally the honey producing capacity of many parts of the country might be greatly improved and future generations of beekeepers be thankful to those responsible.

Reasons why the common lime has fallen into disrepute as an avenue and street tree are: (1) its susceptibility to honeydew and the messiness this implies; (2) early leaf fall; (3) habit of suckering from the base and trunk; (4) large size for suburban streets; and (5) soft nature of the bark of young trees and susceptibility to injury.

The honeydew menace is without doubt the main popular objection to the lime. Not because it may mean spoiled honey, that being the beekeeper's concern and beekeepers are a small section of the community, but because the sticky secretion falls on pavements and streets making them slippery and on anything that may be under them, such as parked cars, public seats, etc. There have been many instances of falls caused by this honeydew, resulting in broken limbs and even litigation. In at least one town known to the writer (Cambridge) lime honeydew has been so bad in some years that the local authority has had sand or gravel strewn on certain streets to render them more safe. In bygone days of slow horse-drawn traffic and pavements and streets not made up or macadamized, honeydew was not the nuisance it is now with fast motor traffic and high grade surfaces.

Early leaf fall causes premature littering of the streets and an autumnal aspect that is objected to on aesthetic grounds. Some limes hold their leaves later in the year than the common lime. The suckering and production of young shoots from the trunk means constant pruning on the part of the local authority. This means labour and therefore expense. It is only the common lime that is a serious offender in this respect. With regard to size and suitability for small suburban streets or grounds of limited extent, there are limes that have a much smaller habit than the common lime.

The following are some of the limes that may be worthy of consideration by those interested in the planting of limes for bee pasturage.

Tilia x euchlora, Crimea lime, a hybrid from eastern Europe introduced to Britain about 1860; of upright growth with dark green glossy leaves and pendulous branches; remarkably free from insect honeydew; flowers and casts its leaves later than the common lime; the flowers are apparently not so freely worked by honey bees as those of some limes; much planted as a street tree in some European countries but not in Britain; probably the most beautiful of limes and an ideal avenue tree.

Tilia mongolica, Mongolian lime; introduced to Britain in 1904; a small tree of slow growth with small maple-like leaves and reddish leaf-stalks; flowers a month later than the common lime; very free flowering at Kew; blossoms small, fragrant and freely worked by bees; the tree is very hardy, of handsome erect appearance, and should be suitable as a small avenue or street tree.

Tilia platyphyllos 'Asplenifolia', cut-leaved or fern-leaved lime. Another small lime where large free-growing trees are not desired; of dense, compact growth with small deeply dissected leaves; exceptionally free flowering; much worked by hive bees; sheds its leaves and flower bracts early.

Tilia maximowicziana, Japanese lime. Introduced to Kew in 1904; a large forest tree in Japan, bears large clusters of flowers which literally hum with bees year after year; at Kew it is more intensively worked by honey bees than any other species.

Other Oriental limes attractive to the hive bee but not yet in general cultivation are: *Tilia insularis* (Korea), *T. oliveri* (central China), *T. miqueliana* (Japan), *T. mandshurica* (Manchuria) and *T. henryana* (central China).

Among the late-flowering limes are: *T. tomentosa* (silver lime), *T. petiolaris* (weeping silver lime), and *T. orbicularis* (a hybrid). These all have a whitish undersurface to the leaves and develop into large trees. If abundant they might considerably extend the lime flow, carrying it well into August. In some years the flowers poison or stupefy a certain number of bees as already pointed out.

The North American limes (*T. americana, T. heterophylla* and *T.h. michauxii*), known as basswood in their native land and the source of much honey, are also good bee plants in Britain but are rarely seen. At Kew they flower two or three weeks later than the common lime.

A study of the accompanying chart of the flowering periods of some of the limes at Kew in a normal season may be of interest

	June	*July*	*August*
Early Street Tree at Kew (*Tilia platyphyllos*)	- - - - - - -	- -	
Cut-leaved Lime (*T. platyphyllos* 'Asplenifolia')		- - - - - - - -	
Common Lime (*T. x europaea*)	————	————	
Japanese Lime (*T. maximowicziana*)		- - - - - - - -	
Crimea Lime (*T. x euchlora*)		- - - - - - -	
Silver Lime (*T. tomentosa*)		- - - - - - -	
Hybrid Lime (*T. orbicularis*)		- - - -	- - - -
Weeping Silver Lime (*T. petiolaris*)		- - - -	- - - -

FLOWERING PERIODS OF LIMES IN AN AVERAGE SEASON AT KEW

The heavy line represents the common lime and the chart illustrates how judicious planting of different species in a district may considerably extend the nectar flow.

The commencement of flowering may vary up to a week or even ten days if the season happens to be an abnormally early or late one. However, the relative periods of flowering of different species of lime remain much the same whether the season be early or late. The chart indicates how the lime nectar flow of an area could be extended from the usual two to three weeks to six weeks were it possible to have a sufficient number of early-, medium-, and late-flowering limes in the same district or locality. Theoretically this should double or even treble the average honey yield as obtained at present where the common lime is the only source. How this could best be achieved the writer leaves to the imagination of the reader. Doubtless the garden city of the beekeeper's dreams has all the streets lined with the smaller limes of one sort or another with a selection of the taller-growing kinds in parks, open spaces and along the borders of playing-fields, providing in all a copious and extended nectar flow.

HEATHER
Calluna vulgaris: Ericaceae

The production of heather honey may be described as the only specialized type of honey production in Great Britain and is largely confined to Scotland (48). The extensive moors in the north of England are also important and a little is produced on the Devonshire moors. The crop is an uncertain one and usually obtained by migratory beekeeping. A really good season only occurs about one year in seven and often no 'surplus' at all is obtained although the hives may become well stocked with stores for the winter. This is often considered by the owner to be sufficient recompense for the cost and trouble of transportation to and from the moors.

Heather or ling is the dominant plant in heathlands in the south of England but these are not important for honey. In eastern Europe the plant covers extensive tracts of country from Brittany to Scandinavia, being typical of poor soils, and in the aggregate is the source of a good deal of the thick strong-flavoured honey so characteristic of this plant. In general the production of heather honey only appeals to beekeepers situated within easy distance of the moors.

Heather is in flower from August, usually mid-August, to about the end of September, and so has a longer flowering period than most honey plants. In the south some flowers may be found open in July. The flowers are smaller than those of the true heaths, with the corolla tubes only 2 to 3 mm. long. Nectar is concealed at the base of the flower and is secreted by eight tiny swellings or nectaries which alternate with the bases of the stamens. As the flower ages nectar secretion ceases and the stamens elongate. The pollen is slate grey in colour and is collected by bees. It is always present in heather honey in great abundance. The grains are in tetrads as in the heaths, somewhat irregular and beset with rows of tubercles.

Conflicting views are held and have been freely expressed regarding the best conditions for the free secretion of nectar in heather. All are agreed on certain points, however, and that the following factors are important: 1, the nature of the soil and subsoil; 2, the age of the heather; 3, rainfall; and 4, possibly altitude. All these factors may of course be discounted by bad bee-flying weather, rain or cold winds, while the heather is out. This often happens owing to the lateness of the season when heather blooms and the fact that it generally grows in areas that are bleak and inhospitable. Actually, the heather flower will secrete nectar at quite a low temperature, probably lower than that required for bees to fly.

In Scotland, J. Tinsley of the West of Scotland Agricultural College, who has made a valuable contribution (22) to the study of heather honey production there, considers the following conditions as the most favourable: 'The type of heather that gives the best yield of nectar is that obtained from young shoots about a foot high. The large bushes of heather, which are several years old, although a mass of bloom, yield very little nectar. On the other hand, young heather which springs up in a year or two after burning of the old heather is rich in nectar. Subsoil plays an important part in the production of nectar. Peat and bog land yield very little nectar, while hill land with a subsoil of granite and ironstone gives the best results. This was particularly noticeable at Leadhills, in Lanarkshire, and judging by the many samples of ling honey received from all parts of Scotland, there is not doubt that the Lanarkshire heather honey cannot be surpassed in quality.'

It has been stated that the honey obtained from ling in the south of England is quite different from that obtained in the north, and that in many parts of the south the plant is of no use at all as a honey producer and does not secrete nectar. The fact that the honey itself may differ from Scottish heather honey is probably because the heathlands in the south are much less extensive and less uniform than those in the north and there is therefore a much greater chance of admixture with nectar or honey from other floral sources. With regard to soil it has been pointed out (Miss Betts, *Bee World*, Nov., 1939) that the nature of the subsoil is vitally important in the south also. Heather is a deep-rooting plant and not a lime lover, and where it occurs in relatively shallow layers of soil overlying limestone or chalk the plants may not secrete nectar at all. On the other hand where the subsoil is an acid sand, as on Bagshot Heath, secretion is good with favourable weather conditions and honey readily obtained, having all the essential characteristics of ling honey.

Sandy, hilly districts are liable to be easily affected by drought in dry seasons, when only the heather in the low-lying areas may yield. In wet seasons the reverse may apply, the plants on the drier well-drained slopes yielding better than those in the low lying boggy situations. It may well be, therefore, that the optimum conditions for heather honey production in the south differ radically from those in the different, more moist, climate of Scotland.

Heavy rain will put a temporary check to the nectar flow in heather, as is the case with many other honey plants. A belief is held in some of the heather districts of Germany and Norway that lightning or thunder will immediately put a stop to the nectar flow in heather and that it will not be resumed until fresh flowers have opened. It is well known that bees are often cross and bad tempered when working on the heather. Can this be due to sudden cessations in the nectar flow not normally apparent to the beekeeper?

Heather honey is considered a bad winter food for bees and prone to cause dysentery owing to its high protein content or the large amount of pollen it contains, which accumulates in the intestine and cannot always be voided during the cold months. This has been definitely proved to be the case in Scotland (22). Many beekeepers in the south of England, however, including the writer, who have made use of heather areas for furnishing winter stores, have not experienced this trouble. Possibly the reason is that in the south, where winters are so much milder than in Scotland, a sufficient number of mild days occur during the average winter to enable bees to take the necessary cleansing flights and so excrete the pollen or pollen remains before they accumulate unduly. The protein content of good ling honey varies from $1\cdot3$ to $1\cdot8$ per cent whereas with ordinary honeys the figure is usually about $0\cdot2$ per cent (30).

Losses of bees in being taken to and from the heather and in failing to locate their new sites on heather have been shown to be greater than is generally supposed. The advantages and increased honey yields likely to be obtained from establishing permanent apiaries on the heather, where this is possible, have also been shown by Tinsley (22) to be very great.

Heather honey has many characteristics which distinguish it from other honeys and place it in a class by itself. By many it is considered the best of all honeys and is much sought after, commanding a higher price than other honey. Some people do not care for it or its strong flavour. This is particularly the case among those unaccustomed to it or always used to the light, mild-flavoured honeys. It is appreciated most as comb honey, the cappings of which are usually white. Being too thick and glutinous (or thixotropic) to be extracted by ordinary rotary extractors, presses are used to obtain run honey. In colour it is some shade of light, dark or reddish brown and numerous air bubbles that are introduced during pressing remain in it and do not rise to the surface. This imparts a distinctive appearance. True heather honey does not granulate but admixture with other honey will cause it to do so, even that of bell heath (*Erica cinerea*) which is usually present on heather moors; more prevalent perhaps on Scottish moors than those in the north of England. The flavour and aroma are very distinct and if a pot of good heather honey be opened in a warm room the aroma can usually soon be detected. It is this strong aroma and flavour that cause some to dislike it.

Heath

The term heath is here used to denote all the true heaths or species of *Erica* as distinct from heather or ling, already dealt with. Five heaths occur wild in the British Isles but only two are widely distributed or common—the so-called purple bell heath or bell heather (*Erica cinerea*) and pink bell heath or cross-leaved heath (*E. tetralix*).

The purple bell heath is a good bee plant and is to be found on moors and heathlands in company with ling. Its degree of prevalence varies. Sometimes it occurs in a more or less pure state, as in Galloway, or only odd patches may be present among the ling. It blooms much earlier than ling and its flowers are a deeper purple, larger and more handsome. They are crowded together in clusters, mainly at the ends of the branches. Their rich hue may be detected miles away and it is to them that heathlands owe much of their beauty. They are a good source of nectar and the bell-shaped corolla is the right length (5 mm.) for the honey bee to negotiate.

In observing the honey bee at work on flowers of bell heath one observer (J. T. Powell, *Journal of Botany L.*, 1884, p. 278) states: 'On July 28th, near Sandringham, I watched for some time the visits of bees. . . . *Apis mellifica* was present in great numbers, and was doing its work in beneficial fashion. I noticed, however, that it also took advantage of holes already pierced in the corollas of very many of the flowers. Both myself and a botanical friend who was with me watched narrowly to see if the honey bee made these holes, but in no case did we observe them do so. We also saw numerous humble bees busy with the same flowers, which they visited in both the above ways.

Presently we saw a *Bombus* pierce a sound corolla, and afterwards several other insects of the same kind repeated the operation. The action was rather boring than biting, and was comparable to pushing an awl without twisting through a thin deal board. In some cases a distinct sound was heard, as when paper is pricked with a pin. . . . The advantage to both bees of the perforation seemed to be that they could sip their sweets in greater comfort in the nearly erect position they assumed during their illegitimate visits than when turned half-over and hanging sideways to insert their tongues into the mouth of the flower; and this comfort seems to be a sufficient motive for the exercise of intelligence in the humble bee.'

Honey is frequently obtained from bell heath either pure or mixed with that of ling. When pure it is of a reddish port wine colour with distinctive or pronounced flavour which resembles somewhat heather honey. Some prefer it to ling but others do not rate it very highly. It is dense but not sufficiently so to prevent removal from the combs with an ordinary extractor as is the case with ling. As the plant flowers so much earlier than ling—it may be in flower in June in the south—it is often possible to obtain it in a pure or relatively pure state. Later on in the year of course it is always blended with that of ling. Bell heath honey is not the best of winter foods for bees, being inclined to granulate in the comb, but does not cause dysentery as ling may do.

Cross-leaved heath (*E. tetralix*) with its leaves characteristically arranged in fours in the form of a cross, and its drooping clusters of pale pink, wax-like flowers, is also common on heathlands. However, it is of doubtful value as a bee plant, the flower-tubes being on the long side (7 to 8 mm.) for honey bees. In the neighbourhood of Bagshot Heath, where the plant is common, the writer has a small out-apiary but has never observed bees working this heath. Flowers with the corolla tubes punctured at the base, however, have been seen.

In parts of Cornwall the Cornish heath (*E. vagans*) is common and beekeepers have obtained surplus from it (*Bee Craft*, Nov., 1935). The honey resembles that of bell heath rather than ling in that it is not thick or jelly-like, i.e. is not thixotropic (30). Irish heath (*E. mediterranea*) occurs wild only in the west of Ireland although common in southern Europe and cultivated in gardens. It is well worked for nectar at Kew in June.

The numerous garden forms of heath that are so much admired

are for the most part good bee plants, having been mainly derived from the wild species already mentioned. It is possible to make a selection so that some will be in flower at all times of the year including the early spring. The majority are much relished by bees. Varieties of *Erica carnea* and *E. darleyensis* which are in flower in March are particularly noticeable in this respect. Beds of the former at Kew, especially 'Springwood White' and 'King George', have been observed on bright sunny days in March covered with bees all feverishly working for nectar. Among the later (August–September) flowering heaths the writer has noticed about three dozen different varieties at Kew and in the fine heath collection at the Royal Horticultural Society's garden at Wisley (Surrey) that are visited by hive bees for nectar. There are doubtless many more. These were mainly forms of *E. cinerea* and *E. vagans*.

The pollen of the heaths, like that of ling, is characteristic, for it is in the form of a tetrad.

FRUIT BLOSSOM

The flowers of all the tree fruits grown in Britain are most useful to the beekeeper for brood rearing if not for surplus honey, their nectar and pollen being available early in the year. Sometimes honey is obtained from them in good districts in favourable seasons. These tree fruits include apple, pear, cherry, plum, quince, medlar, peach and nectarine, the two last mentioned as wall plants as a rule, and only in the milder parts of the country. The only two fruit trees of no avail to the honey bee are the mulberry and the fig. The apple and the cherry are the most valuable as nectar producers.

Soft or berry fruits are a useful minor source of nectar and to a less extent of pollen. They include the raspberry, blackberry, loganberry, various hybrid berries, gooseberry, currants, and strawberries. Of these the raspberry is the best honey producer. (See separate headings in Section 3.)

Apple   *Malus pumila: Rosaceae*

Without question the apple is the most popular and widely cultivated fruit in Britain. This applies also to many other temperate or warm temperate countries. The number of varieties known to cultivation is almost endless and new ones continually appear. They range in size from crab apples and small russets to giant cookers like Rev. Wilks and others. However, it is the apple blossom rather than the fruit itself that is of primary concern to the beekeeper. Few sights can equal an apple orchard in full bloom in the spring such as one sees in Kent and the West Country in cider districts.

It has long been held by British beekeepers that the apple is the best of the tree fruits as a source of nectar and honey, being superior to pear, plum and cherry. It is the last of these fruits to flower, and is therefore more likely to meet the warmer weather as the season advances. Flowering commences some time in April or May, according to season and district, and when stocks are sufficiently strong and weather favourable surplus honey may be obtained. The honey is generally light amber in colour and of good density and flavour. Colour varies with seasons and locality and may be quite dark. The flavour is inclined to be strong at first, but this passes off with age and it remains pleasantly aromatic. It granulates in time, but not rapidly. The great value of the apple flow is to the bees themselves, for not only is it a good stimulus to brood rearing and raising bees for the main flow, but it enables them to store for the lean period of three weeks or so immediately prior to the main honey flow of clover or lime, commencing in the latter part of June. In so many districts there is little nectar coming in at this critical period.

Fortunately there is a good deal of difference in the times when the various varieties of apple commence to flower—as much as three or even four weeks. In most seasons individual trees are in blossom for only about a fortnight, but the fact that several varieties of apple are to be found in most districts ensures a fairly long total flowering period. Although spells of cold or wet weather are not uncommon at this early period of the year, bees are generally able to collect appreciable quantities of nectar for their own use if not for their owner. That the owner's turn does not come round more often than about one year in five seems to be the experience of the rank and file of beekeepers. Those in special fruit districts may do better. With good management, section honey is sometimes obtained from the apple flow.

In forward seasons apples may be in flower at the end of March, but in the case of a late spring there may not be any blossom until the end of April or early May in the same district. Weather conditions in early March, whether warm or cold, have an important bearing on speeding up or retarding flowering. Apple varieties are fairly constant in their flowering periods relative to one another. Among the first varieties to flower are Red Astrachan, St. Edmund's Russet, Irish Peach, Golden Spire, Rev. Wilks, Yellow Newton, Ribston Pippin and Adam's Pearmain. Varieties with mid-season flowering include Stirling Castle, Bramley's Seedling, Peasgood Nonsuch, Allington Pippin, Charles Ross, Lane's Prince Albert and Worcester Pearmain. Among the latest varieties to flower are Crawley Pippin, Royal Jubilee, Mother and King's Acre Pippin. The kind of stock used is known to influence the time of flowering to some extent.

Apple blossoms vary a good deal in colouring and in size. In some varieties the flowers are pure white, as in Christmas-, Claygate-, and Worcester Pearmain; in others pinkish, such as Lord Suffield, Brownlees Russet, Orleans Reinette, and Crawley Beauty; while not a few are richly marked with red or crimson as in Bramley's Seedling, James Grieve, Grenadier, King's Acre Pippin, and Lord Derby.

The apple flower, like that of many other rosaceous fruits, is of the open type and the nectar very liable to be washed out by rain or severely diluted by dew. When dilution reaches a certain point the nectar ceases to be attractive to the honey bee. This may be the reason why bees are sometimes to be seen in the earlier part of the day working apple blossom for pollen only. Later, when the nectar has had time for its sugar concentration to be increased through evaporation and further secretion from the nectary, bees may again be seen working for nectar.

Crab apples and flowering crabs (of the single type) are probably of about the same value as bee plants as are the culinary sorts. They are well worked for both nectar and pollen. The wild crab apple of Britain is often to be seen in oak woods in the south of England.

Apple pollen is produced abundantly and is pale yellow in colour with a slight greenish tinge when packed in the pollen baskets of the worker bee.

Pear   *Pyrus communis: Rosaceae*

The pear is one of the four important tree fruits of Britain, and ranks with the apple, plum and cherry. With them it is of value to the beekeeper as a source of nectar and pollen early in the season for strengthening and building up stocks for the main honey flow. The pear is in general less hardy than the apple and is often grown on walls. It is not grown on the extensive scale that is sometimes the case with the apple. Generally it is to be found in mixed orchards with other fruits. For this reason pear honey is more or less unknown in Britain.

As a nectar or honey producer the pear has always been regarded as inferior to the apple, even in the early or skeppist days of beekeeping. Work on the sugar concentration of nectar or fruit blossom in more recent years confirms this view, for the average concentration of pear nectar has often been found to be considerably lower than that of apples in flower at the same time in the same area. Some varieties of pear in fact have a very low nectar concentration, and there are times when bees visit the blossoms only for pollen, while other fruit trees in the same area are being visited for nectar.

Pear blossoms have a distinctive odour not unlike hawthorn and vary in size and shape in different varieties of pear. In some they are more or less bell-shaped, and in others spread out flat with little protection from rain. It usually takes from five to seven days for all the anthers in a flower to open and liberate their pollen. The pollen is pale or greenish yellow in colour and almost identical with that of the apple and not crimson as stated in some bee books, where confusion has obviously arisen with the unopened reddish-brown anthers!

Most varieties of pear commence to bloom ahead of the apple, but like the apple the date of the appearance of the first pear blossoms varies from year to year according to the earliness or lateness of the season. In most years in the south-eastern districts flowering is at its zenith between the middle and the end of April. The total period of blossoming with pears is not so long as with apples. Some varieties like Vicar of Winkfield and Louis Bonne of Jersey flower early. Others such as Doyenne du Comice, Glou Morceau and Catillac are late flowering. The same order of flowering with different varieties is maintained whatever the season. Individual pear trees remain in blossom from ten to twenty-one days, according to variety and weather during flowering. Low temperatures naturally tend to extend the flowering period (*Journal of Pomology*, 1943, 107–10).

Plum   *Prunus domestica: Rosaceae*

Plums are the hardiest and the most heavy yielding of stone fruits. They are widely cultivated throughout the country. The number of different varieties runs into hundreds. Some are of such long standing in Britain that their source of origin is now unknown. However, a large number of well-known varieties are believed to have been introduced from France at some time or other.

Many plum varieties, probably about a third, are self-sterile, and the question of cross pollination and the insect agents responsible becomes of great importance. Bumble bees and honey bees are normally the most numerous visitors to the blossoms and are believed to account for most of the pollination. Whether bumble bees predominate or not depends upon the surroundings. If these consist mainly of arable land, which is unsuited for the nesting of wild bees, hive bees may be the most frequent visitors, but if rough land such as commons, numerous hedgerows or woods prevail, providing good breeding conditions for bumble bees, they are liable to outnumber the honey bee visitors.

Like the fruits themselves, the blossoms of plums exhibit a good deal of variation in size among different varieties. Some varieties also flower much earlier than others. The cherry plum or red myrobalan (*P. cerasifera*), perhaps hardly a plum in the accepted sense, is the first to flower and is generally ahead of the others by many weeks. Among the true plums, Grand Duke, Monarch, Jefferson, Early Rivers, and Coe's Golden Drop flower early. Late flowering kinds include Belle de Louvain, Kentish Bush, Pond's Seedling, and Gisborne's. Intermediate in flowering are Victoria, Czar, Pershore, and many of the gages. There is generally a difference of about three weeks between the blossoming of the earliest and the late varieties. The duration of flowering is usually two to three weeks or rather more, depending on weather conditions, but varieties vary in this respect also.

In most years early to mid April finds most varieties in flower. Blossoms appear before the leaves and the life of each individual flower, from the opening of the bud until the petals fall, is about a week. The blossoms of some varieties of plum have more scent than others. Hive bees generally show a preference for plum blossom over that of pear and currant, when able to exercise a choice.

The pollen of the plum is similar to that of the cherry, but the individual grains are much larger.

Cherry   *Prunus cerasus: Rosaceae*

Cherries belong to the same genus as the almond and plum, and the flower structure is similar. The honey bee plays an important part in the pollination of the cherry, particularly in the large commercial orchards, where wild bees are rarely sufficiently prevalent so early in the year to carry out the task adequately. In the large cherry orchards of Kent it is a regular practice for owners to make special arrangements with beekeepers to have hives placed among the trees at blossoming time. Skeps of bees were at one time imported from Holland for this purpose.

In return for its services the bee gets both pollen and nectar in plenty from the blossoms when the weather is favourable. As a nectar producer the cherry is considered to be second only to the apple among fruit trees. To obtain surplus honey from the cherry nectar flow, stocks must usually be artificially stimulated to obtain the necessary strength, and weather conditions during the relatively short flowering period are of course vital. Flowering takes place in April or early May as a rule in the south of England.

In most parts of the country cherry trees are only sufficiently abundant to constitute a useful minor nectar source, and to be valuable in building up colony strength for the main flow later in the year. In combination with other fruit trees, however, they help to supply the yields of tree honey that are obtained in so many parts of the country in favourable seasons.

The wild cherries such as the bird cherry (*Prunus padus*) and the gean (*P. avium*) that are so conspicuous in spring in many areas, are of similar value to the honey bee. The bird cherry is common in the north (hagberry of the Scots) while the gean is often abundant in the woods on limestone soils in the south, especially on the Chilterns.

The flowering or Japanese cherries (with single blossoms) now so extensively planted in gardens and as street trees, are similar as bee plants and are well worked for nectar and pollen. One of the earliest is *Prunus x yedoensis* which never fails to attract with its dense masses of snow-white blossom.

Morello cherries differ from eating or sweet cherries in flower characters in addition to the fruit, but are also good sources of nectar and pollen. They flower later and are self-fertile, whereas varieties of sweet cherry are largely self-sterile and require cross pollination with another variety in order to set fruit.

SAINFOIN
Onobrychis viciifolia (formerly *O. sativa*): *Leguminosae*

Sainfoin was introduced to this country as a fodder plant in about the middle of the seventeenth century from the Continent, where it had long been grown. It is still much grown in France and, in fact, in most parts of Europe where chalk soils abound. It is a perennial with a deep penetrating tap-root and so is not very dependent upon surface moisture and able to withstand drought well. As a nectar plant it is reliable and a good source of honey wherever it is grown.

In England the cultivation of sainfoin is largely confined to chalk districts of the south, the three main centres being the Cotswold Hills, Hampshire and adjoining counties, and East Anglia. It is also grown to some extent, alone or in mixtures, in Kent, Sussex, Buckinghamshire and the Vale of Glamorgan. Although the crop is best suited for chalk and will thrive where there is only a thin soil layer over chalk it will grow in other soils provided they are not acid and drainage is good. It is much grown as pasture for sheep and for hay, being one of the best hay crops.

Two main sorts of sainfoin are cultivated, common sainfoin and giant sainfoin, although agriculturalists distinguish three or four varieties of the former. Common sainfoin is the longer-lived plant but usually only flowers once in the season and provides only one 'cut', whereas giant sainfoin, a larger plant, gives two or three 'cuts' and will flower two or three times. It is the better plant from the honey producer's point of view. The best time to cut the plant for hay is about half-way through the flowering period, but often there is delay for one reason or another which is to the beekeeper's advantage if not the farmer's! When grown for seed of course flowering runs its full course. The first flowering of sainfoin usually takes place in the last week in May, the flow lasting about ten days or a fortnight. This is a good time, for it tides the beekeeper over the period between fruit blossom and white clover.

The flowers of sainfoin are a rosy pink colour and a field in full bloom is a pleasing sight. As soon as the first sign of colour appears in a field, that is when the first flower at the bottom of the flower head has opened, bees will be found at work collecting nectar and pollen. Honey bees will neglect other nectar sources as soon as sainfoin becomes available. The flowers secrete nectar freely and will continue to secrete even when temperatures are quite low. The mechanism of the flower and the secretion of nectar is the same as in white and sweet clover.

Sainfoin honey is one of the few honeys that are obtained in Britain in anything like a pure or unmixed state. It is also a very distinctive type of honey, being deep yellow in colour, bright, and with a characteristic flavour and aroma. When freshly gathered in the hive the aroma is not pleasant but this unpleasantness soon disappears (9). The density is not so good as that of white clover as a rule. Most people like the honey and consider it a luxury but some do not care for the flavour. Not a few consider it superior to any other English honey.

The wax of combs built during a sainfoin flow is a beautiful pale yellow in colour. The woodwork of hives and the frames themselves are also prone to become stained yellow. This is attributed to oil in the pollen clinging to particles of propolis on the bees' feet and so getting on to the woodwork, the brownish yellow pollen of sainfoin being of an exceptionally oily nature.

MUSTARD AND CHARLOCK

These two plants, well known to the agriculturist, the one as a farm crop and the other as a weed, are here grouped together for they are of similar value to the beekeeper, yield a similar, characteristic type of honey and are closely allied botanically.

Mustard   *Brassica nigra, Sinapis alba: Cruciferae*

Two distinct kinds of mustard are grown for seed in Britain for the preparation of the national condiment. These are black or brown mustard (*B. nigra*) and white or yellow mustard (*S. alba*), the latter with a larger and paler-coloured seed. Both are equally good as nectar yielders and fields in flower yield a copious supply.

Black mustard needs a deep, moist, fertile soil to produce good crops of seed and is mainly grown in the rich fenlands and alluvial marshlands of East Anglia, where good crops of honey are sometimes obtained from it. Beekeepers in Lincolnshire, Cambridgeshire, Huntingdonshire and Norfolk often benefit from it.

White mustard will succeed under a much wider range of conditions, both as regards soil and climate. Wherever turnips are grown it will succeed. With it there is less loss from shed seed. It is also extensively cultivated as a forage crop for sheep and as a green manure. In the case of both black and white mustard the plants from a single sowing are usually in flower for about a month. March, April or early May is the usual time of sowing for a seed crop.

The honey from mustard is light in colour, bright, and with a strong aroma and flavour when fresh, inclined to leave a slight burning sensation in the mouth (9). Granulation takes place rapidly —more rapidly than with any other English honey. The honey from black mustard is said (23) to be inferior to that from white, being more strongly flavoured and granulating with a coarser grain.

Charlock   *Sinapis arvensis* (formerly *Brassica arvensis*):
                                                              Cruciferae

Although a valuable honey plant, charlock is, at the same time, one of the most troublesome and persistent weeds the farmer has to contend with. Its bright yellow flowers are all too often conspicuous in cornfields throughout the summer months. Besides reducing the yield of corn it harbours the organism responsible for 'finger-and-toe' disease of turnips and other brassica crops. Seed may lie buried in the ground for many years and then suddenly germinate when turned up near the surface by the plough. When old pasture is ploughed up charlock sometimes appears in abundance, the seeds having remained dormant in the soil since it was last farmed as arable. Charlock has in fact been termed a follower of the plough. In some cases it takes command of fallow fields, when the beekeeper's joy becomes the farmer's sorrow. Other local or country names for charlock are karlik, garlock, cadlock, corn mustard, wild mustard, etc.

The yellow flowers are produced in great profusion and are the same in shape and structure as those of other brassicas. Six stamens are present, four long and two short. Nectar is secreted by nectaries at the base of the flower. Sometimes in charlock only those nectaries opposite the short stamens are functional. However, this does not seem to affect the total yield of nectar available to the bee.

The honey from charlock is of excellent quality, being light coloured and of mild flavour. Colour may vary from water white to pale amber. When fresh it may have a faint hotness suggesting mustard, and leaving a mild burning sensation after tasting, but this disappears as the honey ripens. Like honey from mustard the most notable feature about charlock honey is the rapidity with which it granulates. No other honeys in Britain can compare with them in this respect. On exposure to light charlock honey may even granulate within three days (9), and often granulates while still in the hive. For this reason it should always be extracted with little delay. If left in an extractor it may set hard within a few days, to the consternation of the owner when he discovers it.

Its quick setting character is useful when set honey is required at short notice. It is also useful for mixing with other liquid honeys when granulated honey is required. A small quantity added to white clover gives a high grade product, pure white and nearly solid.

HAWTHORN
Crataegus monogyna, C. oxyacantha: Rosaceae

The hawthorn, may or whitethorn is one of our most common native shrubs or small trees. Its strongly-scented white blossoms that appear about the middle of May are familiar to everyone. It is far from being a fastidious plant and will grow in sun or shade and in all soils except acid peat which it avoids. Often it is the only shrub to be seen in pastures, where it sometimes becomes far too prevalent for the farmer's liking. Its virtues as a hedge plant are well recognized and as a farm hedge or fence plant it stands supreme.

As a bee plant the hawthorn is notoriously fickle, being a good source of nectar in some seasons but not in others or in some districts but not others. Attempts to correlate this with soil or with moisture and temperature conditions have not so far met with success and the reasons for this fickleness remain obscure at present. The seasons when hawthorn is a good honey source only come round at long intervals. In some parts of the country 1943 was a good year as were 1911 and 1933 (14).

Sometimes hawthorn blossoms will be worked well and yield honey freely in one area while a mile or two away under apparently similar conditions the blossoms may be deserted by bees. In a district which gives a good hawthorn flow one year the flowers for several succeeding years may offer little attraction. When the flow from hawthorn does occur it is usually very rapid and the smell of the flowers is easily detected in the hives while the nectar is being brought in.

In the hawthorn flower the nectar is secreted by the receptacle or base and is half concealed. In cold or dull weather the inner stamens remain curved inwards but open out in sunshine exposing the nectar more fully.

There are two species of hawthorn, although the differences between them are slight and they seem to be of similar value as bee plants. *Crataegus monogyna* is the more abundant and widespread species, *C. oxyacantha* being confined more to the south-east of the country. The numerous ornamental forms of may with pink or red flowers attract bees when the flowers are single, but not the double forms. So also do the flowers of several introduced species of *Crataegus*, mainly from North America, which are sometimes grown in gardens or as street trees.

Honey from hawthorn is of very high quality. It is usually a dark amber in colour, very thick and of an appetizing rich flavour. Owing to its dark colour and density it has been mistaken for heather honey at honey shows (9). It is not usually bright or sparkling and sometimes has a greenish tinge which detracts from its appearance. The flavour has been described in various ways, although always favourable, such as exquisite, nutty or suggestive of almond.

Usually hawthorn blossoms do not appear until those of apple are over but in some seasons flowering overlaps. The resulting honey, when procurable, which is a blend of apple and hawthorn, is considered to be one of the finest flavoured that could be desired (Tickner Edwardes).

The pale, whitish pollen of hawthorn is freely collected by bees and is often found in honey. The individual grain is similar, microscopically, to that of apple and rose.

SYCAMORE

Acer pseudoplatanus: Aceraceae

Although not really a native of Britain the sycamore is now so widespread, thanks to its wind-borne seeds and the ease with which it propagates itself, as to be considered a wild tree. In some of the beechwoods of the South Downs it is almost as common as the beech itself. The hanging bunches of greenish yellow flowers appear in the latter part of May in most years in the south and only last for two to three weeks, the flowering period being a short one. However, they are a good source of nectar and held in high esteem by beekeepers for their value in stimulating and building up stocks for the main honey flow later in the year.

The sycamore flowers later than its relatives the maples (see Maple). Usually it follows close on the heels of the apple in flowering, but in some years the two flower more or less together, which is then unfortunate from the beekeeper's point of view for they are probably of greater value for stimulative purposes when they flower separately. Surplus honey is obtained from sycamore in some parts of the country in favourable seasons (9). It is usually amber in colour with a greenish tinge, the green colour being possibly due to honeydew which is common on the leaves of this tree. When fresh the flavour of the honey is not of the best, even rank in the estimation of some, but this mellows down with age. Doubtless the flavour as well as the colour is often adversely affected by honeydew. Density is generally fair and granulation slow with a coarse grain.

The flowers of the sycamore, which are about the size of currant blossoms, are arranged in groups of three in long pendent racemes. Usually the centre flower is perfect and develops into the seed while the two lateral ones are male only. They have longer stamens and abortive ovaries. The winged seeds germinate with great facility and seedlings may be seen coming up in all manner of unexpected places.

The sycamore is well suited for cultivation near the sea where it withstands salt sea breezes and maintains an erect position better than almost any other tree. In favourable situations it may attain large dimensions, trees over 100 feet high and 20 feet in girth having been recorded in Britain. There are many garden forms with coloured or variegated leaves.

The sycamore has long been a favourite tree in Scotland, probably on account of its hardiness and rapid growth, and has been much planted around country mansions as well as around farmhouses and cottages on bleak hillsides. In Scotland it is generally known as plane tree. Great maple is another name for it. In the south the tree grows with equal facility and is often planted in windy and exposed situations. Its numerous branches and its large leaves render it a suitable subject where shade is required. It may be seen on farms affording shade for livestock or on the sunny side of farm dairies. As an ornamental tree it is grown either singly or in groups of two or three. It is also effective in avenues, given sufficient room. The litter from its large leaves in autumn is one of its drawbacks, especially in built-up areas. Almost any soil is suited for the sycamore provided drainage is adequate but the tree prefers a dry, free soil to one that is stiff and moist. In spite of its rapid growth it requires several years before it commences to flower and to ripen its seeds.

In many of the wooded mountainous parts of Europe the sycamore is truly wild or indigenous. This includes areas covered by the

Pyrenees, Alps, Carpathians, and the hilly districts that radiate from them.

BLACKBERRY
Rubus fruticosus: Rosaceae

Blackberries are useful bee plants whether they be the common bramble of the hedgerow and field, or one of the cultivated varieties grown in the garden or fruit farm. Flowering takes place late into the summer and provides nectar and pollen at a time when they are becoming scarce in many areas. Wild blackberries generally commence to flower in June or July, reaching a peak in August, but continue until cold weather and frosts arrive. In the case of the cultivated kinds, which may be grown on extensive or orchard lines in fruit districts, the time of flowering depends upon the variety grown—whether early or late.

The wild blackberry grows and flowers freely in almost any soil —whether chalk, poor acid sand, rich alluvium, or clay soils—and never fails to be a source of attraction to the honey bee. It is one of the commonest of wild plants and is to be found in all parts of the country. It is very abundant in hedges, often in company with the dog rose, along roadsides and in fields and meadows. In the scrub of commons and heaths and on waste land it may be very prevalent and form dense clumps up to 10 feet high and more across. It also occurs in the undergrowth of woods but is not so happy in shade and does not flower so freely as when growing in the open.

The common blackberry has a wide distribution. It occurs over nearly the whole of Europe (including Russia) and in central Asia and North Africa but nowhere does it extend to the high altitudes. In the British Isles it is particularly abundant. It has become naturalized in many other countries, even to the extent of becoming a troublesome weed, as in New Zealand.

The flowers of the blackberry are produced on the long main shoots or short lateral ones, generally very freely and over a long period. It is common to see flower buds, flowers, green, half ripe, and ripe fruits present at the same time, which adds greatly to the plant's attractiveness. This long flowering period, which may extend over several months, is largely what makes the plant of such value to the beekeeper.

There are many different forms or varieties of the wild blackberry throughout Britain and various classifications have been put forward for them. The fact that all appear to be useful as bee plants is what primarily concerns the practical beekeeper. Several wild forms or varieties with especially good fruits have been introduced to cultivation from time to time. The flower colour of the wild plants varies, often from district to district, although white is the predominating colour. In some plants the flowers are light pink or tinged with pink, in others they may be pale mauve or lavender. The flowers also vary in their size and in their grouping. However, all are conspicuous by reason of their outspread character. The stamens also spread outwards as they ripen their pollen. Nectar is produced by a circular or disc-like nectary and is easily reached by the honey bee. Other visitors to the flowers include butterflies, wasps, beetles and flies.

In some heathland areas or on the margins of heaths in the south of England where blackberries are a prominent feature of the scrub vegetation and there is no other important honey flow until heather is out, beekeepers consider that what honey they may obtain before the heather flow is mainly blackberry. Such honey is usually of fair quality although pure blackberry honey is not considered to be of the best and to be somewhat coarse in flavour. Pure blackberry honey is probably rarely obtained in Britain, except possibly in the case of hives near extensive blackberry plantations. It is dense and slow to granulate. Whatever the flavour may be this does not detract from its value to the bees as winter stores and it is here that the usefulness of the blackberry to the beekeeper mainly lies. As a honey plant, however, the blackberry is inferior to its close ally the raspberry, which yields a finer honey.

The cultivation of the blackberry on a large or field scale in Britain has only taken place in comparatively recent times, but the fruit is now grown quite extensively for market or for jam making in some districts.

In the earlier days of beekeeping the tough wiry stems of the blackberry, with the thorns removed, were not infrequently used in binding the straw in making skeps. They were also much used in binding thatch.

WILLOW-HERB
Epilobium angustifolium: Onagraceae

The rose-bay willow-herb is one of the best of the wild bee plants of Britain and not infrequently yields surplus honey to beekeepers who are fortunate enough to be in areas where extensive stands of the plant exist. With its spikes of large pink flowers that appear from July onwards it is one of the most conspicuous and handsome of wild plants and a wide expanse of it on the landscape is always a pretty sight. It is frequently among the first plants to appear where fires have been, hence the name fireweed often applied to it. Where heath or woodland fires have taken place it often grows in great profusion. Even on bombed sites in the heart of London it was one of the first plants to appear, the seeds being extremely light and easily carried many miles by the wind. Where woodland has been cut down it may also be very common for a year or two.

Unfortunately for the beekeeper the willow-herb does not remain a dominant plant but becomes ousted by other plants in time. It is therefore of passing value only in most localities. War periods with increased felling of woodland, burning of brushwood, etc., have witnessed a great increase in its prevalence. The name willow-herb is of course due to the resemblance of the plant, before flowering, to a young willow shoot, the leaves being very similar. In parts of Canada the same willow-herb has covered extensive tracts of country in the past, after forest fires or timber felling operations. Large crops of honey have been obtained from such areas. In British Columbia average yields of 100 lb. per hive from willow-herb are not uncommon. So common has the plant been in Canada that in autumn its thistle-like seed down may be everywhere in some areas, even choking up porch screens and meat safes (*Bee World*, 1923, 55).

In Britain the willow-herb is widespread and patches of it are a common sight along roadsides, around rubbish heaps and on the slag heaps of mining and industrial areas where it thrives in a remarkable fashion. Its creeping underground stems, which may ramify up to thirty feet, enable it to increase rapidly apart from seeds and it can become a troublesome weed in gardens, wanting the whole place to itself. The first flowers to open are at the bottom of the flowering spike. By the time those at the top are ready to open seed pods will have ripened at the bottom and other subsidiary flowering shoots have been formed. Thus, flower buds, flowers and seeds may be present on the same plant. It therefore has a long flowering season and is capable of giving a prolonged honey flow when sufficiently abundant. When it occurs on burned-over heathland it is sometimes looked upon as a nuisance by heather beekeepers, as its presence is likely to result in a blended rather than a true heather honey, and it is pure heather honey that always commands the highest price.

The honey from willow-herb is very pale in colour, sometimes water white and of good density but without any very distinctive flavour. Some consider it almost flavourless although very sweet. It is valuable for blending with dark and strong-flavoured honeys. Granulation takes place with a fine grain and is not long delayed as a rule. Wax or comb built while bees are working willow-herb is very pale in colour (14).

Pollen is produced abundantly by willow-herb and is always distinctive when brought in by the bees, owing to its blue colour. The pollen grains are usually bound together by threads of viscin and are fairly large (70 microns).

Other wild species of *Epilobium* attract bees but are not usually sufficiently common to interest the beekeeper. The hairy willow-herb (*E. hirsutum*) is sometimes prevalent in patches by streams and ditches and may be visited for nectar and pollen.

FIELD BEANS
Vicia faba: Leguminosae

The beans in general cultivation in Britain are of four main types: (1) the field or horse bean—a farm crop; (2) the closely allied broad bean of the vegetable garden; (3) the French or kidney bean; and (4) the ubiquitous scarlet runner bean. Other beans have been grown such as the soy bean and the lima bean, but only on experimental lines, not being well suited to the climate.

The field or horse bean (*Vicia faba*), known also as tick or mazagan bean, is far and away the most important to the beekeeper, being so extensively grown throughout the country. Good crops of honey are obtained from it. It is essentially a crop for heavy land and is either spring or autumn sown. Time of flowering depends on sowing and on the variety grown. The size and colour of the flowers also depends upon variety, some having the side petals of the flowers heavily marked with velvety black. In others the flowers are tinged with red. The size of flower may be important in determining to what extent the honey bee is able to work the flowers in the legitimate sense instead of having to rely on the holes made by bumble bees at the base of the flower in order to obtain nectar. This takes place to a large extent in the field bean. The fact that bean pollen is always found in bean honey indicates that the flowers must be worked in the orthodox manner. The flowers are pleasantly scented and a field of beans in full bloom will scent the air on a still day. In good weather honey may be stored in quantity from this crop, and those beekeepers that have access to it in addition to the usual later flows are indeed fortunate. The honey varies from light to dark amber in colour and has a pleasant mild flavour. It is inclined to granulate fairly quickly with a coarse grain.

The broad beans of the vegetable garden are little more than selected forms of the field bean and are of much the same value as bee plants except that they are usually grown on a small scale and, furthermore, often have larger flowers. Extra floral nectaries occur as dark spots on the underside of the leaf bases or stipules. These secrete nectar—produced in sunny but not in dull weather—which is eagerly sought by ants.

Scarlet runner beans (*Phaseolus coccineus*) possess flowers rather too large for the hive bee to negotiate successfully. Here again bumble bee holes at the base of the flower are often available. It is said that when grown as a field crop and the plants pinched back, the size of the flower is reduced and honey bees are able to work them properly. This is stated to occur especially in the Isle of Wight where surplus has been obtained (9).

The French or kidney bean and the haricot beans sometimes grown, have a similar type of flower to the runner bean and are not well adapted to the honey bee. Soy beans are of little or no value as honey plants. This has been proved in recent years in the United States where the crop has been grown over a wide area and has become of great importance. The lima bean, however, which is much grown in warmer climates, may yield good crops of honey.

BUCKWHEAT

Fagopyrum sagittatum (formerly *F. esculentum*): Polygonaceae

Buckwheat is extensively grown as a grain crop in many parts of the world, but in Britain it has never been cultivated to much extent. It is essentially a crop for light land and refuses to thrive on heavy clay soils. The poorest of sandy or light acid soils of the heathland type will often grow a fair crop when all other grains are out of the question. This is the great value of buckwheat, along with its remarkable freedom from pests and diseases. It is even superior to rye in this respect, and accounts for the German name 'heidenkorn' (heath corn) applied to it.

In Britain buckwheat is best known and most frequently seen in Norfolk and Suffolk and the fen areas, where it is usually called brank. It is often used as a catch crop where, for some reason, it has not been possible to get ground ready in time for the intended crop or when a cereal sowing has failed. Its early maturity is then of special value. Buckwheat is often used for game coverts throughout the country to attract pheasants and other game birds. It is also sometimes used in plantations to protect the roots of young trees from drought and for green manuring. During wartime it has been grown more by smallholders as poultry food.

Buckwheat is one of the few plants that offer possibilities of being sown as artificial bee pasturage on an economic basis, the value of the resulting grain crop offsetting cultivation costs. It is in those parts of the country where sandy, acid soils of the heathland type prevail that it is likely to show most promise in this respect. In such areas clovers and lime trees are usually absent and there may be little important bee forage until late in the year when the heather is out. As bee fodder three or four successional sowings at fortnightly intervals, commencing in early May or as soon as the danger of frost is past, is the best procedure. This ensures a long flowering period of two to three months. The young plants are very susceptible to frost and may be killed outright by even one light late frost. Three main varieties of buckwheat are grown, 'Silver Hull', 'Common Grey' and 'Japanese', the last mentioned being the tallest. Sowing is usually done in drills 12 to 18 inches apart.

The value of buckwheat as a honey crop is well known, large quantities of honey being obtained from it in some countries as in the north-eastern United States, Russia and other parts of Europe. It is much grown in Brittany. The honey is always dark and of characteristic strong flavour, not usually appreciated by those unaccustomed to it or used to light, mild-flavoured honeys. It is also generally thick and difficult to extract. Buckwheat honey, however, is always in demand in the confectionery trade and used for special purposes, such as gingerbread, which it helps to keep moist. In London it is popular with the Jewish fraternity, many of whom are familiar with it in Europe. It crystallizes with a coarse grain.

The buckwheat plant is of interest in that two distinct types of flower may be present, although always on different plants. One type has long stamens and short styles and the other short stamens and long styles. Dimorphic flowers of this kind are known in other plants, e.g. the primrose. They assist in promoting cross pollination. Each plant bears only the one form of flower. The white flowers are strongly fragrant and white sepals take the place of petals. Nectar is secreted by eight, sometimes nine, yellow nectaries bound together by a cushion-like swelling at the base of the ovary. The pollen grains in buckwheat are of two sizes according to the type of flower.

Buckwheat does not always secrete nectar freely and the flow seems to be closely governed by the weather. Cool, moist conditions at flowering time are best for good nectar production. It is then that bees work the flowers all day. If the weather is inclined to be hot bees are only attracted in the morning, and if very hot and dry nectar production may cease altogether and bees pay little attention to the plants. In the United States it is known that cool nights and a mean temperature during the blooming period that does not exceed 70° F. provide the best conditions for nectar secretion (see *Beekeeping in the Buckwheat Region*, U.S. Dept. Agric., Farmers' Bull. No. 1216). It has been estimated that an acre of buckwheat is capable of supplying 150 lb. of honey in a season.

DANDELION
Taraxacum officinale (formerly *T. vulgare*): *Compositae*

Some may question whether the dandelion should be included among the major honey plants of the country. However, it is one of the most useful of wild plants to British beekeepers. It occurs everywhere and is regularly visited for nectar and pollen. It is to be found in flower almost throughout the year, but flowers most freely early in the season, before the appearance of fruit blossom, when it is of most value to the beekeeper, especially for brood rearing. It occurs freely in pastures, particularly in chalk districts, when fields may be sheets of yellow at flowering time. This is no uncommon sight on the Cotswolds and Chilterns and dandelion honey has been secured by beekeepers in such areas.

Dandelion honey varies in density and may be deep or pale yellow in colour. It soon crystallizes, doing so with a coarse grain (20). The flavour is strong, particularly when fresh, and the odour reminiscent of the dandelion flower, but this is lessened on ripening. However, those accustomed to mild honeys often do not care for its strong flavour (8).

The pollen of the dandelion is golden yellow but may appear as deep orange in the bees' pollen baskets. It is produced very freely and the bee is able to gather it from the flowers in lumps so to speak. It is somewhat oily, and comb built when dandelion is being freely worked is a distinctive light yellow colour. The individual pollen grain is large and spiny, and is very prevalent in English honey.

There are between 100 and 200 individual flowers or florets in a single dandelion head. The flower tubes vary from 3 to 7 mm. in length and are suitably constructed for the hive bee. At night and in dull or wet weather the heads close up. This protects the pollen and nectar and prevents it from being spoiled by dew or rain. The time of opening of the flowers in the morning varies with the time of year and is much earlier in mid summer than in spring or autumn.

When dandelions occur in quantity in or near large orchards they can be a nuisance to the fruit grower in that bees will often forsake apple or pear blossom in favour of the dandelion flowers, to the detriment of the fruit pollination and subsequent yield of fruit. Whether it is a richer nectar in the dandelion or whether it is the pollen that is preferred does not at present appear to be known.

The dandelion is an important bee plant or honey yielder in many other countries. In Europe it has been listed as an important bee plant for various countries from Spain to Norway. In parts of southern Germany it is a common source of honey. The plant has become widely naturalized in North America and New Zealand where honey is obtained from it and where it is predominantly a valuable spring stimulator.

The so-called Russian dandelions ('Kok Saghyz'—*Taraxacum bicorne* (syn. T. *kok-saghyz*), and 'Krim Saghyz'—*T. megallorhizon*) that have been considered for rubber and grown experimentally in Britain both attract the honey bee freely for nectar and pollen.

OIL SEED RAPE
Brassica napus
(*by* John B. Free, Ph.D., D.Sc.)

The oil seed rape commonly grown in Europe is the swede rape (*Brassica napus*). Its production in southern England has increased greatly in the 1970s and has provided beekeepers with much larger areas of forage for their colonies. Winter crops are sown in late August or early September and harvested in July; spring crops are sown in March and harvested in September. Rape flowers are very attractive to honey bees and, when both winter-sown and spring-sown crops occur in the same locality, flowers may be available from April to July.

The rape flower is hermaphrodite and has four sepals, four petals, an inner whorl of four longer stamens and an outer whorl of two shorter stamens. The anthers of the long stamens, which are about level with the stigma, dehisce inwards. The flower has four nectaries; the two at the base of the short stamens secrete much more nectar, of high sugar concentration, than the two nectaries situated outside the ring of stamens and are much more frequently visited by bees.

All honey bees foraging on oil seed rape collect nectar and none collects pollen only. However, when collecting nectar, they brush against the anthers and their bodies become dusted with pollen grains; some bees pack this pollen into their corbiculae, but others discard it; individual honey bees tend to keep constant to one or other type of behaviour. Each bee visits about 300 flowers per foraging trip and its body is usually covered with over 12,000 pollen grains. The foraging bees may help facilitate pollination and seed yield of some varieties.

Because oil seed rape is such a good source of nectar, many beekeepers move their colonies to rape crops to help increase their honey production. However, any honey produced should be re-removed within a few days, because after that it tends to granulate in the combs, which makes extraction difficult by normal methods.

Insecticides are used on oil seed rape to kill insect pests, in particular the pollen beetle and seed weevil. With an increased area of rape being grown, the number of incidents of honey-bee poisoning associated with insecticide application to rape crops has also increased.

The present level of pests on most crops does not warrant insecticide application, but as more rape is grown pest problems may also become magnified and the necessity for insecticide applications increase. Application of insecticides to flowering rape crops is often unnecessary, but when pest populations do warrant it while the crop is in flower, only those least toxic to bees should be used. When possible they should be applied early or late in the day when bees are not working the crop, or in cool cloudy weather when pollinating insects are not flying.

1978

Section 3

Other Plants Visited by the Honey Bee for Nectar or Pollen

(For ease of reference plants are arranged alphabetically according to their common names. In those cases where there is no common name in general use the scientific (generic) name is used.)

Acacia   *Robinia pseudoacacia: Leguminosae*

The acacia, or more correctly false acacia, also called robinia in France and locust in its native country (eastern U.S.A.), is a well-known ornamental and timber tree. It is of interest in having been one of the first North American trees to be introduced to Britain (63). This took place as long ago as 1640. It has now become established in many other countries, particularly southern Europe, and is a valuable honey plant when sufficiently abundant (42).

In Britain the false acacia is not such a reliable nectar producer as it is in warmer climates or where a hot Continental type of summer is the rule. When warm sunny weather prevails during the flowering period, which usually commences in early or mid June, it is generally well worked and seems to yield nectar freely, but in the absence of hot weather offers little attraction to bees. The bunches of white flowers are delightfully fragrant and each individual flower lasts about a week. Unfortunately, the total flowering period is short, often not more than a fortnight. The tree grows well in poor, sandy soils where many other trees would fail. Growth is very rapid in the early stages and this, combined with its habit of suckering, makes it useful for planting on sandy banks to hold the soil. It is common on railway cuttings and embankments in many parts of France.

The tree yields a good honey, but there are probably few areas in Britain where it would be sufficiently abundant to yield surplus honey. In many parts of Europe, however, surplus is regularly obtained. It is stated (*Bee World*, 1940, 47) to be very common in the Danubian basin in Romania, being freely grown on farms and for timber. It is the source of much honey there which sells at a higher price than other honeys. In Czechoslovakia, Italy and elsewhere in Europe it also affords substantial honey crops. The honey is of a superior type and considered by some to be equal to that of white clover, being light in colour and with good density and flavour. It is slow to granulate. In the United States the tree has been freely planted for timber, to the benefit of beekeepers nearby, but in that country much injury is often done to the trees by borers.

There are many varieties of the false acacia in cultivation, some with variegated or abnormal leaves. The most interesting from the beekeeper's point of view is the everblooming acacia (*Robinia pseudoacacia* 'Semperflorens'), which continues to flower more or less throughout the summer and so lacks the main drawback of the common acacia—a short flowering period. This has been recommended for planting in waste sandy places, mining dumps, etc., on the Continent to improve the honey flow. It has the additional advantage of being thornless.

The nectar in the flower is secreted at the bottom of the staminal tube and the wide calyx allows of its being easily reached by the honey bee. A single flower has been shown to yield 38·4 mg. of nectar during its life with an average sugar concentration of 35 per cent (*Bee World*, 1940, 47). The pollen grains are pale yellow, finely granulated, with three distinct grooves.

The flowers of other species of *Robinia*, e.g. the clammy locust (*R. viscosa*) and the rose acacia (*R. hispida*) have been observed to be visited by honey bees at Kew.

Achillea   *Achillea spp.: Compositae*

The achilleas of the herbaceous border or rock garden are often worked by bees, particularly the stronger-growing kinds, like *A. filipendulina* (Orient), which are best grown in large groups in mixed borders or shrubberies. The flowering period of these perennials is usually from July to September.

Aconite   *Aconitum napellus: Ranunculaceae*

The common aconite or monkshood is mentioned as a bee plant by some writers, but probably in error, the winter aconite (*Eranthis hyemalis*) being intended. See Winter Aconite.

Agrimony   *Agrimonia eupatoria: Rosaceae*

This wild British plant is sometimes visited by bees for pollen, but does not yield nectar. Its familiar yellow flowers have a faint odour of lemon and appear in June and July, a time when there is an abundance of pollen available from other plants. If the plant flowered early or late in the season it would doubtless be more freely visited for pollen.

Alder   *Alnus glutinosa: Betulaceae*

Alder is one of the early-flowering woodland trees bearing catkins, like willow and hazel, that may be a useful source of fresh pollen for early brood rearing, provided the weather is warm enough for bees to fly and trees are relatively near the hives. The catkins may open in February or March. In very early seasons they may even open at the end of January. The tree is widely distributed in Britain, especially near streams and in damp situations. In the wetter parts of the country it is often abundant in oak woods. Other European and American alders are also useful for pollen.

Allium   *Allium spp.: Alliaceae*

The alliums of the flower garden, which may have mauve, crimson, red or yellow flowers are for the most part visited by bees for nectar. Were it not for the strong onion-like odour when crushed they would probably be more popular and generally grown. The genus *Allium* also includes important vegetable crops such as the onion (*A. cepa*), leek (*A. ampeloprasum porrum*), chive (*A. schoenoprasum*), all good nectar plants, also many wild species such as wild garlic and ramsons (*A. ursinum*) which honey bees freely visit and which are sometimes common in meadows and woods. In some areas their presence in the herbage may be a nuisance by causing tainting in milk.

Crops like onions and leeks are of course normally harvested before they flower, but when they are grown on a field scale for seed, honey may be obtained during the blossoming period. Onion honey, which has been obtained on seed farms in California and elsewhere, has been described as amber in colour with an onion flavour which disappears as soon as the honey is fully ripened (23). Probably that of other alliaceous plants would be similar. The handsome mauve flowers of chives which are borne in such profusion in early summer are worked most assiduously by the honey bee both for nectar and pollen. In the case of the leek the large flower-heads may contain from 2–3,000 flowers and each head must therefore contribute an appreciable amount of nectar.

The actual secretion of nectar in an *Allium* flower is of interest. It may be secreted at the base of the ovary or by three double septal glands situated half-way down the ovary. Nectar is produced very freely under favourable conditions.

Almond   *Prunus dulcis* (formerly *P. amygdalus*): *Rosaceae*

In Britain the almond is grown solely as an ornamental or flowering tree and not for nuts as in warmer climates. The masses of pink blossoms that appear so early in the spring are always a pleasing sight to the beekeeper who knows their value to his bees for nectar and pollen after the long winter months of confinement. At this early season much will depend on weather conditions and whether day temperatures are high enough for bees to fly. Fortunately there are marked differences in the times when flowering commences with the different trees, usually to be found in streets and gardens, and this extends the total flowering period. Blossoms may be available for three weeks to a month. Usually a number of fine or warm days occur during this period, allowing bees to visit them. In some years the flowers may be severely damaged by frost.

In the south of England flowering may take place any time from mid February (as in the early spring of 1943) to early April, depending on the conditions and the variety of almond grown, that known as 'Praecox' being one of the earliest. The flowering almonds belong to the bitter almond group for the most part and have darker coloured flowers than the sweet almonds, which are grown for nuts. In these the flowers are generally pale pink or almost white.

Nectar secretion in the almond flower is very profuse under suitably warm conditions. This may be demonstrated by placing a twig of almond blossom in a warm room overnight with the cut end in water and a bell jar or inverted vessel of some sort over it to maintain a still humid atmosphere. The next day the base of the flower will be observed to be swimming in nectar. The nectar is first secreted in small droplets on the brown inner surface of the cup-shaped 'receptacle'. These increase in size if not sipped away by bees or other insects and eventually coalesce so that the base of the flower is flooded with nectar. The almond also possesses extra floral nectaries on the leaves like the cherry and cherry laurel, which are visited mainly by ants and wasps.

Alyssum   *Lobularia maritima: Cruciferae*

Sweet alyssum or madwort, as it was called in olden times, both a wild and a garden plant in Britain, is much favoured by bees, its white honey-scented flowers appearing in July and August. In the West Country it is often to be seen on the tops of walls and in dry sandy places, where it grows as a perennial. The many garden forms with flowers of various colours are grown as annuals. Another well-known alyssum is that called 'Gold Dust' (*A. saxatile*). It is a native of southern Russia and one of the most popular of yellow spring flowers. Bees visit the flowers although not zealously, and probably mainly for pollen.

Anchusa   *Anchusa spp.: Boraginaceae*

Some good bee plants exist among the anchusas, both wild and cultivated. Their blue or purple flowers are worked predominantly for nectar. *A. azurea* is perhaps the commonest and best garden species, reaching 3 to 5 feet. If not allowed to seed this perennial will flower continuously from June to September. The variety 'Dropmore' is one of the most handsome. The Cape anchusa (*A. capensis*), a much smaller plant but equally favoured by bees, may be grown as an annual or biennial.

Alkanet (*A. officinalis*) and evergreen alkanet (*Pentaglottis sempervirens*), both natives of southern Europe, are found wild or naturalized in many parts of Britain. They are good nectar plants and are recommended for the wild garden. The latter is quite common on and around rubbish heaps at Kew, its sky-blue flowers appearing in spring.

In alkanet and other anchusas the nectar is secreted by the four-lobed base of the ovary and is concealed in the flower-tube by hairs near the entrance. This protects it from rain and from short-tongued insects. As the tube is about 7 mm. long in alkanet the nectar is just within reach of the honey bee.

Anemone   *Anemone spp.: Ranunculaceae*

The white flowers of the well-known wood anemone (*A. nemorosa*) appear in early spring, sometimes as early as the middle of March, and may be a useful source of pollen. The flowers yield a pale-coloured pollen in abundance and honey bees may be frequent visitors. It has been stated in regard to this plant that 'the honey bee not only collects pollen but also sucks, boring with its proboscis into the base of the flower, so as to obtain the sap which it requires for moistening the pollen' (11).

Other anemones may be visited for pollen such as the pasque-flower (*Pulsatilla vulgaris*), wild on chalk downs as well as cultivated, also the autumn-flowering Japanese anemones.

Anise Hyssop *Agastache anisata* (formerly *A. anethiodora*): *Labiatae*

This name has been used for a North American plant of the mint family that has attracted a good deal of attention as a bee plant in recent years. It bears heads of pretty mauve flowers and dark green leaves. There have been many references to it in the American bee press (*American Bee Journal*, 1943, 454).

The plant was first grown in Britain in 1826, but failed to take on as a garden plant at that time. Little was heard of it until the last few years when it was 'rediscovered' in the United States and also grown by some interested persons in the British Isles, including the writer. It is said to have been widely distributed at one time from Lake Superior and Manitoba to Nebraska westward, and to have been used as a beverage plant and for seasoning by Indian tribes, in the same way that we might use sage. Some of the earlier settlers reported fine crops of honey from the plant, the honey possessing in some slight degree the same fragrance. The flowers are undoubtedly very attractive to honey bees and bumble bees for nectar. It ranks high as a bee plant in America and has the advantage of a long flowering season—from June until frosts arrive. Being a perennial and easy to grow and propagate, it is worth a place in the bee garden. As a general garden plant, however, its apparently straggling habit under British conditions may be against it. Furthermore the plants the writer has seen and handled have had little or no fragrance in the leaves. This is strange but may be due to the different climatic conditions in Britain.

Other closely-related species are known to be good nectar plants, especially the giant hyssop of California (*A. urticifolia*), which yields crops of a light-coloured, minty-flavoured honey, slow to granulate.

Anoda   *Anoda cristata* (formerly *A. hastata*): *Malvaceae*

At Kew this mallow-like plant, which is a native of Mexico, is freely worked in the late summer for pollen. It is not in general cultivation.

Anthericum   *Anthericum spp.*: *Liliaceae*

These tuberous-rooted plants are sometimes seen in gardens and are visited by bees. The white flowers of the St. Bernard lily (*A. liliago*), appearing in June and July, and of *A. ramosum*, may be utilized for both nectar and pollen.

Antirrhinum   *Antirrhinum majus*: *Scrophulariaceae*

The common antirrhinum or snapdragon of the flower garden is essentially a bumble bee rather than a honey bee flower. Only the former is sufficiently powerful to open the mouth of the flower and reach the nectar. However, honey bees and other small bees are sometimes able to enter faded flowers and probably secure a certain amount of nectar. They are also able to get nectar from holes made by other insects near the base of the flower-tube when these are present. They may also collect pollen. Nectar is secreted by the front part of the base of the ovary which is smooth and fleshy.

Apricot   *Prunus armeniaca*: *Rosaceae*

As this fruit is best suited for cultivation under glass or against a south wall in Britain, it is nowhere sufficiently abundant to be of importance to the beekeeper. It flowers early and is visited for nectar and pollen and so may be helpful for early brood rearing. In warmer climates where orchards of the trees exist the apricot is considered to be a valuable bee plant and a heavy nectar yielder in suitable weather.

Arabis   *Arabis spp.*: *Cruciferae*

Among the several species of *Arabis* found in gardens the alpine rock cress (*A. alpina*) is probably the most useful to the beekeeper, flowering as it does so early in the year (March) and yielding an abundance of pollen and nectar. It is common in gardens and rockeries everywhere although sometimes misnamed white alyssum or alison. Few plants yield nectar so freely so early in the year and are so much appreciated by bees. The name 'honigschub' (honey bush) is used in Dutch for these plants. The flower of alpine rock cress has two pairs of nectaries at its base and the nectar collects in dilations of the sepals immediately beneath them.

Among the wild species of *Arabis* the hairy rock cress (*A. hirsuta*), which is sometimes common in dry rocky places, is also worked for nectar and pollen.

Aralia   *Aralia spp.: Araliaceae*

The two aralias sometimes grown for their ornamental foliage and the oddity of their thick club-like branches appear to be good bee plants. They are the angelica tree (*A. elata*), a prickly shrub or small tree originating in Japan, and the Hercules club (*A. spinosa*) a native of the south-eastern U.S.A. and very similar. The masses of small white flowers arranged in huge bunches or panicles up to 2 feet in length have been observed to be covered with bees in August and September at Kew. The flowers are very fragrant and the nectar, which must be secreted abundantly, is clearly visible.

Arnica   *Arnica spp.: Compositae*

Hardy, dwarf, herbaceous perennials, allied to senecio but not often cultivated. The yellow flowers of *A. montana* (mountain tobacco), a good rock garden plant, are visited by bees. Both the flowers and root or creeping rhizome of this Central European plant have medicinal uses, so it is just possible that it may be cultivated more extensively at some future time.

Artichoke   *Cynara scolymus: Compositae*

The globe artichoke, when allowed to flower for seed production, is visited freely by bees, but the nature of the honey is not known. Normally of course the flower-heads are cut for use before they open.

The Jerusalem artichoke (*Helianthus tuberosus*), a native of North America in spite of its name, is also a good bee plant in countries where it flowers. In Britain it does not normally flower.

Ash   *Fraxinus excelsior: Oleaceae*

The small and inconspicuous flowers of this well-known tree, which are devoid of both sepals and petals, are borne in small clusters on the branches. They appear early in the spring and are known to be visited by bees for pollen, but are probably of little importance.

Asparagus   *Asparagus officinalis: Liliaceae*

The flowers of the common asparagus of the vegetable garden or asparagus fern of the flower bed, are much sought after by both bumble and honey bees for nectar and pollen. They are pendulous and bell-shaped with a characteristic odour, the male flowers being conspicuously larger than the female. The plant occurs in its wild form in some coastal districts of Britain and is known to have become naturalized in many parts of the world, seeds being distributed by birds. Honey has been obtained from it which in France has been described as greenish and of mediocre quality (4) and in California as amber or dark coloured, of lower market value than many other honeys (23).

Aster   *Aster spp.: Compositae*

A number of popular garden plants belong botanically to the genus *Aster*, including the perennial asters or Michaelmas daisies. The genus contains over 400 species and most are probably useful to the honey bee, having a suitable flower structure. At Kew numerous species of *Aster* not in general cultivation have been observed to be freely worked. Most of the cultivated asters and Michaelmas daisies are native to North America where they often occur in large masses in the wild state. Along with the wild golden rods (*Solidago*) they afford good bee pasturage, especially in the autumn months. The honey obtained has been variously described but whatever its flavour it affords useful winter stores.

The Michaelmas daisies, now so much cultivated for late summer or autumn flowering, and largely of garden or hybrid origin, all appear to be freely worked for both nectar and pollen. They are undoubtedly very useful to urban and suburban beekeepers as late minor food sources, especially pollen, in areas where they are freely grown. Some of the North American asters are now naturalized in Britain.

The common annual garden aster so much used for bedding is not a true *Aster* in the botanical sense, the many varieties being derived from a Chinese plant (*Callistephus chinensis*). The single forms are visited by bees but not so freely as the Michaelmas daisies.

The name Michaelmas daisy is, confusingly, sometimes used for the sea aster (*A. tripolium*), a wild British plant found near the sea and often brightening the salt marshes from July to September with its yellow-centred blue flowers. It occurs also in Europe and a Dutch writer states 'where there is salt soil on the coasts there are often masses of sea aster and much honey is obtained from it' (*Bee World*, 1927, 48).

Aubrieta   *Aubrieta deltoidea: Cruciferae*

Being among the first of our common garden plants to flower in the early spring, and by reason of their producing such masses of bloom, aubrietas are useful to the beekeeper. They supply their visitors with both nectar and pollen, the nectar being freely produced at the base of the flower and collecting in large drops in the cup-shaped bases of the sepals. The original plants were native to mountainous regions of southern Europe, but there are now numerous garden forms and colour varieties in cultivation.

Azalea   *Rhododendron spp.: Ericaceae*

Azaleas, like rhododendrons, are not of special merit as bee plants, and are little visited by the hive bee. This is probably because, in most cases, the nectar is too deep seated to be readily available to the honey bee, although often much sought after by bumble bees. However, in the case of the dwarf small-flowered azaleas, flowering in early spring, the writer has observed the blossoms being freely worked for nectar by the honey bee when the weather has been favourable.

Balsam   *Impatiens spp.: Balsaminaceae*

The cultivated garden balsams (*I. balsamina*) grown as hardy and half-hardy annuals are mostly 'double' and of no use to the honey bee. An interesting Himalayan balsam (*I. glandulifera*) which is often cultivated and is now naturalized in some parts of the country, appears to be a good bee plant. It reaches four feet or more in height and has large pink or white flowers with a wide mouth. Bees walk into the flower, disappearing out of sight, and draw the nectar from the narrow curved spur at the base of the flower. As this is only some 5 mm. in length all the nectar is available to the honey bee. In entering and leaving a flower the bee's back is dusted white with the pollen from the overhead stamens. Another Himalayan balsam favoured by bees and usually in flower in August and September is *I. amphorata*.

Baptisia   *Baptisia spp.: Leguminosae*

A vigorous group of perennials from North America, not unlike lupins. False indigo (*B. australis*), which has indigo blue flowers, is sometimes seen in cultivation and attracts bees.

Barberry   *Berberis vulgaris: Berberidaceae*

The wild barberry occurs in some parts of the country, especially on chalk. This prickly deciduous shrub bears clusters of small yellow flowers in May which are attractive to bees. The nectar is secreted by thickened tissue or nectaries at the base of each petal and collects between the bases of the stamens and the carpels. The pale yellow pollen is discharged by the interesting movements which the stamens exhibit on being touched.

Many other species of *Berberis* ere cultivated as ornamental shrubs, and are worked by bees for nectar or pollen, the two best known being *B. stenophylla* and *B. darwinii*. Both these are evergreens and the former is much grown as a hedge plant, producing an abundance of blossom.

Bartonia   *Mentzelia lindleyi (syn. Bartonia aurea): Loasaceae*

This Californian annual has been described as a good bee plant (9). It reaches 1 or 2 feet in height and bears showy golden flowers that are pleasantly fragrant. Sowing should be carried out in groups or patches in warm, open situations.

Basil   *Ocimum spp.: Labiatae*

The flowers of both sweet basil (*O. basilicum*) and bush basil (*O. minimum*) are visited by bees. These aromatic culinary herbs are much less cultivated or used in Britain than they are in Continental countries, but are worth a place in the herb garden. There are many varieties differing mainly in the size, shape and colour of the leaves.

Bearberry   *Arctostaphylos uva-ursi: Ericaceae*

This small trailing undershrub which occurs abundantly in heaths in Scotland and the north of England, as well as in the colder parts of America and Europe, is sometimes cultivated. It is especially useful for covering unsightly objects and as a ground cover. Clusters of rose-coloured flowers are produced in May and June. These contain nectar and are visited by honey bees, the nectar being secreted by a fleshy ring surrounding the ovary. Dense hairs protect and hold it at the base of the flower.

Bedstraw *Galium spp.: Rubiaceae*

Lady's bedstraw (*G. verum*), a very common wild plant with small yellow flowers, has been referred to as a bee plant (27), but it is doubtful whether any of the bedstraws are of any real consequence to the honey bee.

Bee Balm *Melissa officinalis: Labiatae*

The name bee balm is sometimes applied to the red bergamot (*Monarda didyma*) which also has sweet-scented leaves. This may be misleading to some. The true bee balm of our forefathers is a native of the Mediterranean region, but is common in English gardens. It naturalizes freely and is almost a weed in some instances. The white flowers do not, in the writer's experience, attract bees to any extent in spite of the assertions in some of the older bee books. The flowers of the plants examined have too long a corolla tube for the honey bee to extract all the nectar, and it is unable to force its head into the widened part of the tube.

However, the real interest of bee balm to the beekeeper is in the aromatic, lemon-scented leaves. If these are crushed and rubbed inside a skep it is said to render it attractive to bees, and that if convenient branches of trees near an apiary be rubbed with the herb at swarming time, swarms will settle there. Rubbing the hands with the leaves is claimed to, and probably does, help in preventing stings. The scented water in which leaves have been macerated and soaked has been used in sprinkling bees in uniting—before the general adoption of the newspaper method. The fragrant oil from the leaves, known as Melissa oil and used in perfumery, has been recommended for making scented syrup for introducing queens.

There is difference of opinion as to the value of bee balm in attracting swarms. Most of those who have tried it in Britain consider it quite ineffective. A Continental observer has stated that while he knew it to be quite effective in Jugoslavia when he lived there, he found it useless in Germany (*Bee World*, 1930). Possibly in cool and humid climates it is not as effective as in warm countries.

Beech *Fagus sylvatica: Fagaceae*

The common beech does not flower freely every year in Britain, but bees are known to visit the flowers for pollen. As flowering does not usually take place until May, when many other sources of pollen are available, it is doubtful whether it is ever of much consequence. However, bees have been known to collect beech pollen freely in Europe (*Bee World*, 1936, 87). In some years in Britain the beech may be a troublesome source of honeydew.

Begonia   *Begonia spp.: Begoniaceae*

The dwarf begonias often used as bedding plants may be visited a good deal for pollen, which is yielded freely.

Bergamot   *Monarda spp.: Labiatae*

The name bergamot or bee balm is used for garden species of this North American genus, on account of the aromatic sweet-smelling leaves, especially *M. didyma* and *M. fistulosa* with red and purple flowers respectively. The brightly-coloured flowers are too long for honey bees, but bees are sometimes to be seen obtaining nectar from holes at the base of the flower made by bumble bees. In Texas and elsewhere in the United States, however, other species with shorter corolla tubes, especially *M. punctata*, commonly called horse mint, are first-class bee plants and important sources of honey.

Bidens   *Bidens spp.: Compositae*

Mainly North American annuals, but rarely cultivated. Some are known to be good honey plants in their native land and are usually termed Spanish needles or sticktights on account of their burr-like seed heads. *B. pilosa*, now widely spread as a weed in warm climates is well worked for nectar at Kew. There are two wild British species usually found in damp places.

Bilberry   *Vaccinium myrtillus: Ericaceae*

The bilberry, also called blaeberry, huckleberry, whortleberry or whinberry in different parts of the country, is often abundant on moors and in heathy mountainous areas. It is best known for its edible fruits and rarely exceeds 2 feet in height, with wax-like drooping flowers that appear from April to June. These secrete nectar freely and are relished by bees. Generally the plants are in out-of-the-way places and not many bees are on the moors when they are in flower. The flowers of the less common mountain blaeberry (*V. uliginosum*) and cowberry (*V. vitis-idaea*) which appear later, are also worked for nectar.

Some of the North American vacciniums are known to be good bee plants, especially the tree huckleberry (*V. arboreum*), one of the main sources of honey in Arkansas (19). This small tree will grow in Britain, but is slow growing.

Birch *Betula spp.: Betulaceae*

The birches are not important as bee plants, but yield an abundance of pale yellow pollen early in the year which is sometimes collected. The silver birch (*B. pendula*) is common in the south of England, especially in the sandy soil of heath land. In the north and west *B. pubescens* is the more common.

Bird's-foot Trefoil *Lotus corniculatus: Leguminosae*

This small, clover-like plant is common in dry pastures, often in association with white clover, and its heads of yellow flowers are conspicuous from June onwards, these being followed by pods arranged like a bird's foot—hence the name. It occurs freely wild and is also a common constituent in seed mixtures for permanent pasture, being often particularly suitable for soils that are unsuited for red clover. Possessing a deep tap-root it is able to thrive in poor dry soils and is very drought resistant. It is frequently to be seen in great abundance in chalk soils and pastures near the sea, where it is usually in flower about a fortnight before white clover. At some of the higher altitudes it may be the only plant of the clover family in the pastures. The writer has noticed bees working the flowers in July, in Dorset, in preference to those of white clover growing round about it. Being deep rooted it may well be less fickle than white clover as a nectar yielder.

The value of bird's-foot trefoil as a nectar plant in Britain is probably not fully appreciated, although it is held in high esteem in some countries, e.g. Switzerland. There are some who consider the plant has not received the attention it merits from the agriculturist. A drawback from the farmer's point of view is that stock do not seem to like the plant when it is in flower. However, it is much grown for pasture or fodder in parts of Europe. In central France it is even grown as a self crop, in the same way as lucerne, such areas being called 'lotières'. Several different varieties are recognized such as broad-and narrow-leaved and indigenous.

The marsh bird's-foot trefoil (*L. majus* or *L. uliginosus*) is a similar plant that grows in moist situations. Bees visit the flowers also. The name trefoil is applied to various other clovers or clover-like plants—species of *Trifolium* and *Medicago*.

Blackthorn   *Prunus spinosa: Rosaceae*

All lovers of wild flowers welcome the snowy-white blossoms of the blackthorn or sloe, which contrast so well with its black, leafless boughs. Appearing as early as March or February in some years, they are among the first harbingers of spring. The shrub is common in woods, coppices and hedges throughout Britain, but not in northern Scotland, and flourishes in a variety of soils. It often forms extensive local patches which arise from its free suckering habit. The flowers are good sources of nectar and pollen in country districts, when weather conditions allow bees to fly. To the rural beekeeper it is in a way the counterpart of the almond to those with bees in towns.

The bullace and the wild plum, so similar to the sloe, are also similar in their flowering characters and value as bee plants.

Bladder Senna   *Colutea arborescens: Leguminosae*

The yellow pea-like flowers are sometimes visited by honey bees, the probosces being inserted laterally for nectar, which means of course that pollination is not effected by them. This Mediterranean shrub is much grown in gardens and has become naturalized in some places. It is often to be seen on embankments of railways and new arterial roads.

Blood Root   *Sanguinaria canadensis: Papaveraceae*

The reddish-yellow juice of this pretty Canadian plant is what accounts for its name. It is sometimes grown in gardens for its white flowers which appear in early spring. It also had medicinal properties. Like other members of the poppy family it is much visited by bees for pollen. It is a hardy plant that soon spreads and is well adapted for the semi-wild garden.

Bluebell
Endymion non-scriptus (formerly *Scilla nonscripta*): *Liliaceae*

From time to time the question has been raised as to whether the common bluebell, so abundant in many parts of the country, is a useful bee plant. The corolla or flower-tube is obviously too long for the honey bee to collect nectar in the ordinary way, but there seems to be some evidence that it is able to get at the nectar from the side of the flower near the base (9). The pollen, which is a very pale shade of blue, is also collected. Some observers testify to having seen bluebells being very freely worked by bees (*Bee World*, 1928, 148) but this does not appear to be general.

Many of the cultivated scillas, which have smaller flowers than the wild bluebell, are well worked for nectar and pollen in the early spring, and are valuable to the beekeeper where they are much planted. The Siberian squill (*S. siberica*), and its varieties, is among the most popular and earliest to flower. Nectar is secreted by the septal glands of the ovary and collects between this organ and the bases of the stamens. The pollen is bluish-grey in colour and conspicuous in the bees' pollen baskets as it is brought in at the hive entrance in urban areas where scillas abound. The Siberian squill is useful for edging and for naturalizing on lawns, provided the grass is not cut until the plants have matured. In some seasons it is in flower as early as February and supplies much-needed pollen when few other sources are available.

Bog Asphodel   *Narthecium ossifragum: Liliaceae*

The star-like yellow flowers of this plant, which is often abundant in bogs and moors, may be visited by bees for pollen. Flowering generally takes place in July and August.

Bogbean   *Menyanthes trifoliata: Gentianaceae*

Many consider the bog- or buck-bean one of the most beautiful of native plants. It is to be found only in boggy places and on the edges of pools. Bees sometimes visit the flowers, mainly for pollen.

Borage   *Borago officinalis: Boraginaceae*

Borage is one of those plants well established in the esteem of beekeepers and frequently grown by them in order to watch the bees at work on the attractive sky-blue flowers. Of European origin it has been grown in Britain for many centuries, the flowers and leaves being a favourite ingredient in several beverages, especially claret cup. Two or three leaves impart a refreshing flavour resembling cucumber—now often used in place of borage. Flowers were also used for garnishing salads. The plant is sometimes to be found in waste places and cultivated ground, and when once established in a garden will come up year after year from self-sown seed. It has become naturalized in other countries, including parts of Australia.

The plant is easy to grow and succeeds in most soils. It is best planted 1½ to 2 feet apart to allow of free development. Flowering commences in the middle of summer and continues until cold weather or until the plants are cut down by frost. It is often grown purely as an ornamental plant and violet-red and white-flowered varieties exist. For early flowering, seed may be sown in the autumn in many parts of the country.

The nodding flowers of borage yield nectar freely and are sometimes to be seen humming with bees, including bumble and solitary bees. Owing to their usually inverted position the nectar is not easily washed out by rain. They are similar in this respect to the flowers of the raspberry. Each flower has a black cone of anthers in its centre. The nectar is secreted by the receptacle at the base of the ovary and collects between, and is concealed by, the bases of the stamens. To obtain the nectar the bee simply hangs under the flower and inserts its proboscis between the stamens. In doing this, pollen becomes sprinkled on its body.

It is probably rarely that borage honey has been obtained in anything approaching a pure form and this may account for the conflicting descriptions of it. One well-known authority describes it as 'excellent' (20) while another report states: 'In *Practicher Wegweiser*, page 280, Herr Willhelm says, that in response to the general cry, "Sow Borage," he has been sowing it for years and now has it in abundance! But, alas! now that he has it in such abundance that it shows its character in the surplus honey, he finds it such as no customer wants, and says it is as black as a certain "gentleman" with whom beekeepers do not generally care to have dealings. The task of getting it now rooted out is a difficult one'! (*American Bee Journal*, 1908, 103.)

Bees will work borage freely as a rule, and all day long, but sometimes forsake it for other plants such as lime and white clover when these are in flower and yielding well.

The pollen of borage is a light bluish grey or almost white, and the individual grains as seen under the microscope have the constriction in the middle, characteristic of many of the borage family (*Boraginaceae*).

Box   *Buxus sempervirens: Buxaceae*

This well-known evergreen tree is much more abundant in parts of
Europe and Asia than Britain, where it favours the chalk districts of
the south, a good example being Boxhill in Surrey. The inconspicuous
yellowish-green flowers appear early in the year (March to May)
and are sometimes, although not always, well worked by honey bees.
The flowers yield both nectar and pollen, but it is probably the latter
that is the chief attraction. It is yellowish green in colour and is
produced abundantly.

The flowers are of two kinds, an apical female flower being surrounded by six or more male flowers. Both types of flower produce a
small quantity of nectar. The action of the honey bee in collecting
pollen has been described by Müller as follows: 'It frees the pollen
from the still undehisced anthers with its mandibles, regurgitates
some honey from its slightly protruded proboscis and then transfers
the pollen by means of the front and mid legs to the hind ones. All
this, however, is done so quickly that the individual acts can
scarcely be followed' (11).

Honey from box has been described as of indifferent quality (27).
If this is so it is not surprising as other members of the same family
are known to cause bitterness in honey. In parts of South Africa
honey is not infrequently spoiled through bees obtaining nectar from
the wild tree euphorbias (noors honey).

Bracken   *Pteridium aquilinum: Polypodiaceae*

Although bracken, like other ferns, bears no flowers, bees have been
observed visiting it. Extra-floral nectaries occur on the leaf stalks and
these seem to attract the honey bee at times, when there is a dearth of
other sources of nectar. However, it is rarely that bees pay any
attention to bracken.

Brassica   *Brassica spp.: Cruciferae*

A large number of brassicas afford good bee pasturage when in
flower. Many of these are everyday vegetables such as cabbages,
cauliflowers, broccoli, Brussels sprouts, kales, turnips and swedes.
Others are less known, such as kohl-rabi and Chinese cabbage,
while still others, which are farm crops such as mustard and rape,
are more important as bee fodder on account of their being grown
on a larger scale and maintained until flowering is completed. Some

wild plants and weeds such as sea-cabbage and charlock, also good
bee plants, are included in the genus *Brassica*. (See charlock, mustard,
radish, turnip, oil seed rape, etc.)

Most of the brassicas of the vegetable garden are harvested before
the flowering stage is reached, but quite often unused plants or
crops are left and flower before being removed. The flowers of all
attract bees and in the case of large-scale planting for seed as on seed
farms, honey may be obtained. This is basically the same as honey
from charlock and mustard with the same tendency to rapid
granulation.

The flowers of all brassicas are very similar, being mostly yellow
with four petals and sepals instead of five, which is usual in most
plants. Nectar and pollen are usually produced freely, the nectar
being secreted at the base of the flower and collecting there or in the
cavities formed by the curved sepals.

Broom   *Cytisus scoparius: Leguminosae*

In company with gorse and heather, broom is one of the common
shrubs of moorland and sandy commons, often growing vigorously
on bleak rocky hillsides and reaching 5 to 6 feet in height. Its rich
golden flowers appear in May and June and are frequently visited by
honey bees. There is difference of opinion as to whether bees obtain
nectar from the flowers, although they are known to collect pollen,
which is deep orange when packed in the bees' pollen baskets.

Some of the cultivated or ornamental brooms, particularly the
smaller-flowered sorts, have been observed by the writer to be well
worked by bees for both nectar and pollen. Notable among these is
the early-flowering Portuguese broom (*C. albus*) which bears white
flowers, and is in bloom before most of the other sorts.

Bryony   *Bryonia dioica: Cucurbitaceae*

The rather unattractive greenish-white flowers of white bryony are
visited by bees. They may be seen from May to September in hedgerows and other places for this climbing plant has a long flowering
season. Although pollen is collected, the visits of hive bees often
seem to be primarily for nectar, especially in the latter part of the
summer. The nectar, which is partly concealed, is secreted at the
bottom of the naked fleshy cup formed by the fusion of the lower part
of the calyx and corolla.

Buckthorn   *Rhamnus catharticus: Rhamnaceae*

This spreading shrub of woods and hedgerows is sometimes very prevalent on chalk hills in the south of England. Its small greenish-yellow flowers appear in dense clusters in May and are eagerly visited by bees for nectar. As the flowers are of a simple open type the nectar is exposed and available therefore to many short-tongued insects which must compete with the hive bee for it.

An allied shrub, the alder buckthorn or dogwood (*Frangula alnus*, formerly *R. frangula*), less dense of habit, grows mainly on damp acid soils. Like the buckthorn it was once cultivated to supply charcoal for gunpowder. Its greenish-white flowers are visited for nectar and pollen in the early summer.

Many other introduced species of *Rhamnus* have been observed to be well worked for nectar at Kew. One of the most interesting of these is *R. purshiana* from western North America, the bark of which is the source of the drug cascara. It yields surplus honey in its native land. The cultivation of this tree on extensive lines for bark has been considered in Britain and elsewhere, so its potential value as a honey producer is of more than passing interest to the beekeeper.

Buddleia   *Buddleia spp.: Loganiaceae*

The common blue and purple buddleias seen in gardens and so popular with butterflies are of no use to the honey bee as the flowers are far too long. However, the yellow buddleia (*B. globosa*) from Chile and Peru with its flowers in attractive yellow balls is well worked for nectar and appears to be a good bee plant. Flowering usually takes place in early June, a dearth period in many areas, and the arrangement of the flowers in small spheres is such that a bee alighting on one of the heads is able to thrust its proboscis into the separate flower tubes one after another in an incredibly short space of time.

The shrub is evergreen in mild districts and reaches 15 feet. It is easily grown and distinctive among buddleias with its balls of bright yellow, sweet-scented flowers arranged in clusters.

Buffalo Berry   *Shepherdia argentea: Elaeagnaceae*

This North American shrub, allied to our sea buckthorn, is not in general cultivation but has been referred to on the Continent as an early-flowering shrub of value to beekeepers (*Die deutsche Beienen-*

zucht, Nov., 1934). The flowers are produced in clusters in March
from the joints of the previous year's growth. As an ornamental
shrub it is of little value.

Burdock   *Arctium lappa: Compositae*

As one of the commonest weeds this plant needs no introduction,
its burr-like seed heads, that cling so tenaciously to man and beast
alike, being so familiar. The purple flowers that appear from July
onwards closely resemble those of thistles. They are much frequented
by hive bees which may be observed collecting the white pollen and
probing the flowers as for nectar.

As the corolla tube of the individual flowers generally exceeds
8 mm. in length it may be that the honey bee is only able to get
nectar when it is secreted freely and rises in the base of the tube
to within reach of the bee's tongue. As bumble bees—also frequent
visitors—with their longer tongues would be able to draw the nectar
at any time, it is doubtful whether the ordinary burdock can be of
much value for nectar to hive bees where bumble bees are plentiful.
These remarks do not apply to the small burdock (*A. minus*) in which
the whole plant, including leaves and flower-heads, is smaller.

Burnet   *Sanguisorba spp.* (formerly *Poterium*): *Rosaceae*

These meadow plants are sometimes attractive to bees for pollen, the
lesser or salad burnet *S. minor* (formerly *Poterium sanguisorba*)
being prevalent on chalk soils. They were at one time planted as
pasturage. The seed is still used sometimes in seed mixtures for
permanent pastures or poor, light or thin soils.

Butterbur   *Petasites hybridus* (formerly *P. vulgaris*): *Compositae*

This plant with downy leaves like coltsfoot, but larger, is often to be
seen along river banks and in other situations. The flower stalks
appear early, from March onwards, and before the leaves. Separate
male and female flowers generally grow on different plants. They are
attractive to bees, although not very conspicuous in themselves. One
observer states: 'Where it abounds I feel convinced that it is of great
help to bees, blooming as it does so early in the year, for I have had
evidence that it yields abundantly, both nectar and pollen. There is a
bed of it not more than half a mile distant from Nottingham Castle,
which, on warm days in March and April, is literally alive with hive

bees, even to the abandonment of patches of Coltsfoot growing near it' (8).

This plant is said to have been planted by Swedish beekeepers near their hives on account of its early flowering. For those who may wish to do likewise it is well to remember that the plant has long creeping roots with which it multiplies quickly, and that it may oust other plants or become a nuisance if not controlled.

A closely allied plant is the so-called winter heliotrope (*P. fragrans*), very like coltsfoot in leaf, and a rampant weed. It bears spikes of dingy lilac flowers from December to February. Bees visit the flowers when the weather is suitable. Like the butterbur it will take command of a garden if allowed to do so.

Buttercup   *Ranunculus spp.: Ranunculaceae*

The numerous buttercups so prevalent in pastures are of little consequence as bee plants. The flowers of many species seem never to be visited by honey bees at all but those of others, e.g. the lesser celandine (*R. ficaria*) and bulbous buttercup (*R. bulbosus*) both common species, may be worked for pollen on occasions.

Buttercups are, in general, unpalatable plants, owing to the presence of an acrid poisonous principle, and have caused poisoning in livestock. It is of interest, therefore, to note that in recent years the pollen of buttercups has been proved to be actually injurious to bees in Switzerland and elsewhere in Europe and responsible for a form of May sickness. Bad outbreaks of this malady have occurred in seasons when cold weather has retarded the flowering of the usual early pollen plants, like cherries and dandelions, but not the more hardy buttercups, causing the bees to collect larger amounts of buttercup pollen.

In Britain there is usually an abundance of other wholesome pollen plants in flower at the times when buttercups are in bloom, so presumably this form of bee malady may be less likely to occur. The harmful nature of buttercup pollen, or at any rate that of some species of *Ranunculus*, may be the reason why the flowers are so often completely neglected by hive bees. Their instinct warns them to leave the flowers alone. (*Bee World*, 1942, 47, 78.)

Calendula   *Calendula officinalis: Compositae*

Calendulas or marigolds enjoy wide popularity and are easily grown, being not at all particular as to soil and surroundings. Unfortunately

for the beekeeper most of those cultivated, including the choice
varieties, are 'double' and of no use for bees. Sometimes reversions to
the single form take place, however, when bees freely work the central
or disc florets. The old-fashioned single yellow or cottage marigold
is a good bee plant.

Californian Poppy   *Eschscholtzia spp.: Papaveraceae*

The Eschscholtzias or Californian poppies are always popular for
mixed borders and thrive equally well on light and heavy soils.
Besides the original yellow and orange shades they are now available
in other colours, some of the pale rose and flesh coloured shades
being particularly delicate.

These plants are always favoured by bees for pollen, which must
be of a kind particularly to their liking. It is bright orange in colour.
The flowers may sometimes be worked for nectar also, possibly
secreted only spasmodically or in small amounts.

Callicarpa   *Callicarpa bodinieri giraldi* (formerly *C. giraldiana*):
Verbenaceae

This attractive shrub is sometimes to be seen in parks and pleasure
grounds, but is a comparatively recent introduction from central and
western China. The flowers, which appear in July, are decidedly
attractive to bees for nectar. Handsome foliage and blue berries are
additional merits of the plant.

Camassia   *Camassia spp.: Liliaceae*

These hardy bulbous North American plants are often grown in
gardens, especially for cutting. Their handsome, usually blue,
flowers are attractive to bees, particularly those of *C. cusukii* and
C. scilloides (quamash, formerly *C. osculenta*). The bulbs of the
latter were used as food by American Indians.

Campanula   *Campanula spp.: Campanulaceae*

Campanulas are invariably popular bee plants and are visited for
nectar and pollen. The number of different kinds cultivated is very
large and they range from tiny Alpine plants (harebells), a few inches
in height, to tall, vigorous-growing perennials like the chimney
bellflower (*C. pyramidalis*) 4 to 6 feet high. Canterbury bells in their
many and richly-coloured forms are included with them.

Among the better known garden campanulas the Carpathian

harebell (*C. carpatica*), popular for rock gardens, has been observed to be freely worked on occasions, so also has the peach-leaved bellflower (*C. persicifolia*). Many campanulas are good edging plants, and when grown in this way provide a greater quantity of blossom for bees to work upon.

The wild campanulas of the woods and downs are also visited, but some of the species are comparatively rare plants. The clustered bellflower (*C. glomerata*) may be frequent in dry hilly pastures and attracts bees. The harebell or bluebell of Scotland (*C. rotundifolia*) is also a common species. Campanula pollen is often found in heather honey where plants occur on the moors.

Candytuft   *Iberis spp.: Cruciferae*

Both annual and perennial forms of candytuft are much cultivated. The annual kinds, which flourish in almost any soil are probably the more favoured by bees. They are now available in numerous colours besides white, including purple, lilac, crimson, rose and carmine. For early flowering seed may safely be sown in the autumn in most districts.

Cardoon   *Cynara cardunculus: Compositae*

The flower-heads of this vegetable which are so like those of the artichoke, are visited by honey bees. However, the plant is seldom seen in Britain although more freely cultivated on the Continent. The leaves are blanched and used like celery.

Carrot   *Daucus carota: Umbelliferae*

The flowers of carrots when left for seed are visited by honey bees along with numerous other insects. The nectar is freely exposed as in most umbelliferous flowers, consequently short-tongued insects like flies have free access to it and probably compete actively with the honey bee. This may explain why it is not as attractive to bees as are many other flowers where the nectar is partly concealed and available to fewer insects.

Carrot plots or small fields grown at Kew during World War II were not well worked by hive bees even when in full flower during fine weather, although occasional bees were present and not interested only in pollen. However, in California, honey is said to be obtained from carrots when grown for seed and to be of light amber colour (23).

The wild carrot, progenitor of the cultivated sorts, and with similar flowers, is a very common plant in dry places and the borders of fields. Honey bees have been observed on it. The plant has become a common weed in parts of North America. It is listed as a honey plant there (19).

Caryopteris   *Caryopteris spp.: Verbenaceae*

Little-known semi-woody shrubs from China and Japan, some of which are inclined to be tender except in the mildest parts of the country. *C. tangutica* is probably the hardiest. *C. incana* is popular in gardens. The violet-blue flowers appear late (September to October) in dense clusters and are much visited by hive bees for nectar at Kew. The leaves are sweet scented.

Castor Oil Plant   *Ricinus communis: Euphorbiaceae*

An important oil seed crop in warm countries. Sometimes grown as an ornamental foliage plant in Britain. Bees may visit the flowers for pollen which is produced freely. Extrafloral nectaries exist at the base of the leaf.

Catalpa   *Catalpa bignonioides: Bignoniaceae*

The catalpa or Indian bean from eastern North America has been grown in English gardens for about 200 years, and is popular on account of its small size and elegant shape, its attractive flowers, and its quaint bean-like seed pods. The blossoms appear in July and August in large panicles or bunches and are available when few other trees are in flower. They attract hive and bumble bees in large numbers for nectar. As the individual flowers are bell-shaped and about $1\frac{1}{2}$ inches long and across bees crawl right into them. Bees have been observed collecting nectar from the extrafloral nectaries on the undersurface of the leaves. These secrete before, during and after the blossoming period (*American Bee Journal*, 1938, 319).

Other catalpas not in general cultivation have been observed to be popular with bees at Kew. Notable among them is *C. ovata*, a Chinese tree with smaller flowers and more deeply lobed leaves than the common sort. It flowers equally freely and at about the same time.

Catmint   *Nepeta spp.: Labiatae*

The name catmint is applied to several different plants. In the days of our forefathers it was always used for an erect-growing wild British plant bearing white flowers dotted with pink (see Catnip, *Nepeta cataria*). The name has also been used for calamint (*Calamintha spp.*), but is more generally applied nowadays to *mussinii* (formerly *N. marifolia*), the common garden catmint that originated from south-eastern Europe and Persia.

This is one of the most widely cultivated perennials and always an attractive plant with its dense prostrate habit, its grey-green foliage and blue flowers that appear in such profusion in May or June. It remains in flower for a long period and attracts bees continually. Nectar is secreted freely and the flower tubes are just the right length for the honey bee and sufficiently long to exclude flies and other short-tongued insects. It is undoubtedly one of the best bee plants as any beekeeper who has observed it closely can testify. Unfortunately, it is never cultivated on a sufficiently large scale for surplus honey to be obtained—having no other use than as a garden plant. It grows well in almost any soil, even thriving on the gravel covering of air-raid shelters during the war.

The flowers of several other allied species of *Nepeta* that closely resemble the common catmint are also favourites with the honey bee.

Catnip   *Nepeta cataria: Labiatae*

As already stated, this herb is also known as catmint. Its white flowers are spotted with purple or pink and are attractive to bees. Although a native plant it is not very common in the English countryside except sometimes in chalk soils, where it may occur fairly freely in hedgerows, reaching 2 to 3 feet in height. It is strongly scented and the odour seems to be attractive to cats—hence the name.

Early settlers from Europe introduced the herb into North America, where it has become widely naturalized and is regarded as a good bee plant. Moses Quinby, the famous American beekeeper, is said to have stated that if he were to grow any plant intensively for the honey it produces that plant would be catnip (20).

Catsear   *Hypochoeris radicata: Compositae*

The yellow flower-heads of this weed, which somewhat resemble those of a dandelion, are visited by bees—probably mainly for pollen.

Ceanothus   *Ceanothus spp.: Rhamnaceae*

Most of these blue-flowered shrubs in cultivation are of hybrid origin. The flowers are visited by bees, but not in large numbers as a rule, and mainly for pollen. Those that flower early in the year (April) seem to offer most attraction. The shrubs are generally inclined to be tender and are best with the protection of a wall. In their native home (North America) they occur in great abundance in some regions, yielding nectar and pollen. They constitute the wild lilac of the chaparral.

Celery   *Apium graveolens: Umbelliferae*

The flowers of celergy attract honey bees to some extent in company with many other insects, and where this plant has been grown in bulk for seed—as in the seed belt of California—surplus has been obtained (*Gleanings in Bee Culture*, 1919, 712).

Wild celery, the ancestral form of the cultivated plant, grows about ditches and rivers and in moist places generally. The clusters of small white flowers appear from June to September.

Celtis   *Celtis occidentalis: Ulmaceae*

A large North American tree sometimes to be seen in gardens and pleasure grounds although not often. Like its close relative the elm, it may be a useful early pollen source. The name sugarberry is sometimes applied to it.

Centaurea   *Centaurea spp.: Compositae*

Included in this genus are a number of useful bee plants which occur wild, as weeds, or are cultivated as garden plants.

One of the most common is knapweed or hardheads (*C. nigra*) a valuable nectar plant and the source of surplus honey in Ireland. The honey is said to be golden, thin and with a sharp flavour—not of the best quality in fact—and the pollen is greenish-yellow. The plant is common everywhere in fields, meadows, roadsides, etc., and flowers freely in June and August.

The greater knapweed (*C. scabiosa*) is also a good bee plant. Clumps of it are sometimes conspicuous with their bright purple flowers on sea cliffs. The brilliant blue cornflower (*C. cyanus*) which may be noticeable in ripening corn is also much sought after by bees as are the garden forms of the plant often grown as annuals.

With these bees appear to work the blue, white, or many-coloured forms with equal fervour and do not discriminate between them. Cornflower honey on the Continent has been described as viscid and of a golden colour (*Bienen Zeitung*, July 1926).

Other centaureas which are particularly attractive to bees are the bluebottle or mountain centaurea (*C. montana*), a more robust plant and useful for cutting, *Centaurea dealbata* from the Caucasus, and sweet sultan.

The so-called yellow star thistle (*Centaurea solstitialis*), a Mediterranean plant now widely spread as a weed (including Britain) is also a good plant for bees, and surplus honey has been obtained from it where it is common enough. It has become a serious weed in grain fields and along roads and railways in parts of the United States, where it is said to provide a slow but continuous nectar flow and yield a honey of fine flavour, slow to granulate, and much in demand (23).

Cephalanthus   *Cephalanthus occidentalis: Rubiaceae*

This shrub is widely spread in Canada and the United States where it is an important honey plant and commonly called button bush as the small white flowers are in globular heads. It grows well in Britain, but is not often seen, thriving best in moist situations and preferably peaty soils. The fact that it flowers fairly late (August), when few nectar sources are available in some districts, makes it of interest to the British beekeeper. Honey from the plant in its native land is said to be light in colour and mild in flavour (19).

Chamomile   *Anthemis spp.: Compositae*

These daisy-like flowers, particularly those of the corn chamomile (*A. arvensis*), may be visited by bees in June and July for nectar, but usually there are other more attractive sources of nectar available at this time.

Chestnut   *Castanea sativa: Fagaceae*

The sweet or Spanish chestnut is regarded as having been introduced to Britain by the Romans and is now common everywhere, except on chalk soils which it avoids. Flowering usually takes place in July and the flowers are of two kinds, the tassels of male flowers being more conspicuous than the female. Bees are frequently to be observed assiduously working the male flowers for pollen, but little nectar seems to be obtained as a rule, at least under the average conditions prevailing in Britain.

However, in parts of Europe where the tree is often grown extensively for the sake of the nuts, it seems to be of some importance as a nectar plant and honey is obtained from it. This is the case in Spain (*Flora Apicola de España*, M. Pons Fabreques) and in parts of southern Switzerland (*Bee World*, 1941, 95). The honey is reputed to be yellow in colour and of rather poor quality. In Czechoslovakia nectar is said to be secreted only after warm nights. The reason why so little is heard of the tree as a nectar yielder in Britain may be that the climate is not warm enough or is in other respects unsuitable. In Japan chestnuts are important bee forage.

Chickweed   *Stellaria media: Caryophyllaceae*

As one of the commonest weeds of the garden and rich cultivated land, this plant needs no introduction. It may be found in flower from February to October. Bees have been observed visiting the flowers for nectar, which is only secreted in sunny weather. The five tiny nectaries are situated at the base of the five outer stamens. In California the chickweeds are regarded as valuable for early nectar for stimulative purposes (23).

Chicory   *Cichorium intybus: Compositae*

This beautiful plant is one with many uses. It may be grown as a crop for the roots (for coffee), as a garden vegetable for the young leaves, or as a forage plant for stock. It also occurs as a weed in waste places and on the borders of fields, especially in light gravelly soils. The sky-blue star-like flowers, as large as a dandelion, appear from June onwards and are great favourites with honey bees. They supply nectar and pollen and may be found to close up early in the afternoon.

Chicory has been grown on a field scale for the roots and for seed in Huntingdonshire, where honey has been obtained from it. This is described as of a peculiar yellow colour, slightly greenish, even when granulated, and with a flavour reminiscent of chicory when fresh.

Cistus   *Cistus spp.: Cistaceae*

These shrubs have mainly originated from the Mediterranean basin and are generally referred to as rock roses—along with *Helianthemum*. Most of the garden forms are of hybrid origin. Their brightly-coloured flowers, which are usually present in great profusion, attract bees—mainly for pollen. The flowers seldom last more than a day, in some cases only for a morning. The fact that many cannot withstand severe winters accounts for their not being more generally cultivated.

Clarkia   *Clarkia elegans: Oenotheraceae*

The flowers of this hardy annual, when of the single type, receive the attention of honey bees. The double-flowered forms, which include many of the choice newer varieties, have little or nothing to offer bees and fail to attract them.

Claytonia   *Claytonia spp.: Portulacaceae*

The claytonias are native to northern Asia and North America, but two species, *C. perfoliata* and *C. sibirica*, have long been naturalized in Britain. The first mentioned, originally introduced from north-west America as a pot-herb, is now a very common annual weed in some parts of the country. It favours moist places, growing 6 to 12 inches high and bearing white flowers.

Among those grown in gardens are some with pretty rose-pink flowers such as *C. virginica*. Those that blossom early in the year are attractive to bees, especially for pollen.

Clematis   *Clematis spp.: Ranunculaceae*

These popular climbing plants produce an abundance of pollen from the many stamens of their flowers. It is often collected by hive bees. Some, but not all, yield nectar as well. Many of the showy garden forms are among those without nectar, including the widely grown 'Jackmanii' hybrids.

The wild clematis or traveller's joy (*C. vitalba*) of the hedgerows yields nectar in addition to pollen, and is sometimes buzzing with bees. It flowers in midsummer, when other more important nectar plants are generally available, consequently nothing is known regarding its honey (14). The flower is of interest in that the nectar is produced in droplets on the filaments (stamen stalks) and not from nectaries at the base of the flower.

Cleome   *Cleome spp.: Capparidaceae*

Some interesting bee plants are included in this genus although none is in general cultivation in English gardens. Several are essentially hot-house plants in this climate, but others have been successfully grown in the south of England as half-hardy annuals, cultivation being best in light dry soils and warm situations. They bear large showy flowers with long thin or spidery stamens, hence the name spider flower sometimes given to them.

Among those worthy of consideration by enthusiastic beekeepers and for the bee garden are the Rocky Mountain bee plant (*C. serrulata*) (19) and *C. lutea*. Both are annuals which have attracted a good deal of attention among American beekeepers (*American Bee Journal*, 1940).

Clethra   *Clethra alnifolia: Clethraceae*

This handsome shrub is the most interesting of the clethras from the beekeeper's point of view. It was introduced into Britain from North America over 200 years ago. Like so many of the heath family, it thrives best in a peaty soil and requires a moist situation. It reaches 8 to 9 feet in height and produces fragrant white flowers in August which are of value to bees. In its native land, where it is called sweet pepper bush, it is regarded as a good nectar plant yielding a thick, white honey of fine flavour (19). It has proved to be attractive to bees in Germany (*Bee World*, 1935, 120).

Colletia   *Colletia armata: Rhamnaceae*

This peculiar Chilean shrub, which reaches 10 feet in height and bears long cylindrical spines up to 1½ inches long, flowers late in the year (October). Its small, white blossoms are attractive to bees on sunny autumnal days at Kew—nectar and pollen being collected. It is little grown at present, but may have possibilities as a hedge or fence plant.

Collinsia   *Collinsia spp.: Scrophulariaceae*

The brightly-coloured flowers of these annuals are sometimes visited by bees. Their early flowering, especially when autumn sown, is one of their main virtues, particularly from the beekeeper's point of view.

Coltsfoot   *Tussilago farfara: Compositae*

As one of the first wild flowers to appear in early spring and a source of pollen and nectar, this humble little plant is a good friend of the beekeeper. It is common in fields and pastures, especially on clay soils and marls, and is one of the first plants to appear anywhere where ground has been disturbed. Railway tracks and the cuttings and embankments of some of the newer arterial roads may be covered with the plant. The felted leaves were much used medicinally at one time and are still collected in quantity for the manufacture of substitute tobaccos or herbal smoking mixtures.

The flowers generally spring up early in March, before the leaves, and are not unlike dandelions, but smaller. They close up at night and on dull days. Pollen is produced by the central male flowers of the flower-head and nectar by the outer female ones, a yellow circular nectary being at the base of each style (11). The pollen is golden in colour and the individual grains densely spiny when magnified, like other members of the same family.

As a bee plant the value of coltsfoot is primarily for pollen, weather conditions being seldom favourable enough for free nectar secretion at the early season when it flowers. It is doubtful whether the flowers are ever as freely worked as those of the dandelion are on occasions.

Columbine   *Aquilegia vulgaris: Ranunculaceae*

The wild columbine occurs fairly frequently in some parts of the country, notably Devon. The spurs of the flower which contain the nectar are too long for the honey bee, although it sometimes steals the nectar from the punctures of bumble bees. The flowers may also be visited for pollen. This applies also to the garden aquilegias.

Comfrey   *Symphytum spp.*

Wild and prickly comfrey have been listed as bee plants for nectar and pollen (*Bee Craft*, May, 1936). With most comfreys, however, nectar can only be reached by the honey bee when the base of the long flower-tube has been punctured by a bumble bee.

Coneflower   *Rudbeckia spp.: Compositae*

These American plants, very like sunflowers with their large yellow heads but with a raised central cone, appear to be good bee plants. They flower in the late summer and autumn and when common in gardens may, along with other autumn flowers, assist in supplying winter stores. *R. laciniata*, with deeply-divided leaves, is one of the tallest, growing from 7 to 10 feet in height. It has a large flower with a greenish cone and is sometimes freely worked by bees for nectar.

Convolvulus   *Convolvulus arvensis: Convolvulaceae*

As one of the most troublesome of farm weeds, the common convolvulus or field bindweed is well known to all who live in country
districts. It may also be a nuisance in the garden of the town dweller
or on the allotment because the smallest fragment of the underground stem or root left in the soil may give rise to a new plant. The
funnel-shaped pink or white flowers which are sometimes striped
vary a good deal in size. Nectar is secreted at the base of the ovary
and the flower has long tubular nectar passages leading to it. The
smaller-flowered kinds are visited freely by honey bees, while bumble
bees work both the small and the large-flowered sorts equally well.
Bindweed which had heavily infested cornfields in the Penn (Buckinghamshire) area was observed by the writer to be intensively worked
by honey bees for nectar in July, and appreciable quantities of nectar
must be obtained from such fields. Bindweed pollen is frequently
found in honey, especially in honey from Europe (27).

Other bindweeds (species of *Convolvulus* and *Pharbitis*, formerly
Ipomoea) are known to be honey yielders. One of the best known is
perhaps the campanilla of Cuba (*Pharbitis sidaefolia* or *P. triloba*)
which is of great importance as a honey plant to the beekeepers
there. The honey obtained from it is said to be equal to that of
lucerne or sage in flavour and colour, and the comb built from it a
pearly white, yielding a wax as white as tallow (20).

Coreopsis   *Coreopsis spp.: Compositae*

Both the annual and perennial forms of the garden coreopsis or
calliopsis receive attention from honey bees for nectar and pollen.
Although the plants may be in flower for a long period in the
summer they are seldom cultivated in anything but small groups or
patches. Many of the true species, as distinct from the garden forms,
have been observed at Kew to be very freely worked in the late
summer.

Corn Cockle   *Agrostemma githago* (formerly *Lychnis githago*):
Caryophyllaceae

The showy purple flowers of this plant, sometimes conspicuous in
corn as it reaches the ripening stage, may attract the attention of
bees. The pollen has often been found in honey, the individual grain
being large and somewhat unusual in structure (27).

Cotoneaster   *Cotoneaster spp.: Rosaceae*

This large group of deciduous and evergreen shrubs contains some
first-class nectar plants. About two dozen different species have been
recorded as well worked by honey bees at Kew. Few flowers are so
persistently visited as those of some of the commonly cultivated
cotoneasters. Even in the middle of the lime flow certain cotoneasters
in flower have been observed covered with bees which suggests that
the nectar may be of high sugar content or at any rate especially
attractive to the honey bee. Nectar is obviously secreted very
copiously in many instances. This may be seen by placing sprigs of
blossom in water overnight in a warm, close atmosphere, when the
bases of the flowers will be found to be covered in nectar by morning.
Secretion takes place from the fleshy inner wall of the receptable as
in the almond and other *Rosaceae*.

The species most commonly cultivated in Britain as decorative
shrubs, largely for their attractive berries and in some instances as
hedges, are probably *C. simonsii*, *C. microphyllus*, *C. frigidus* and *C.
horizontalis*. These are mostly from the Himalayas. The two first
mentioned have become quite extensively naturalized in some parts of
the country, seed being carried by birds. *C. frigidus* develops into a
small tree while *C. horizontalis*, as the name indicates, is low growing
and well suited for training on walls or wooden fences. The last
mentioned is invariably besieged by bees when in flower.

Cotoneasters are amongst the easiest subjects to grow in the
garden or shrubbery and thrive in any soil. The flowers are remarkably
uniform, generally about ¼ inch across and well constructed for the
honey bee. Flowering takes place mainly in May and June. Several
are in flower in the first half of June, a dearth period for nectar in
many areas.

Cow-wheat   *Melampyrum pratense: Scrophulariaceae*

Hive bees are sometimes to be seen around the pale yellow flowers
of this very common wild plant. They cannot obtain the nectar in
the orthodox manner as the flower-tube is much too long for them,
being some 14 to 15 mm. in length. However, the obliging punctures
of other insects at the base of the flower are common and are the
cause of the attraction. Nectar is produced freely at the base of the
flower and may rise 2 to 3 mm. in the flower-tube. An interesting

E

point about the plant is that it bears nectar secreting trichomes or
hairs that attract ants.

Cranesbill   *Geranium pratense: Geraniaceae*

The common purple-flowered meadow cranesbill, or wild geranium,
that is often so common around thickets and in damp places, is a
good bee plant. So also are several other wild and garden geraniums,
including the bloody or blood-red cranesbill (*G. sanguinum*), doves-
foot cranesbill (*G. columbinum*) and dusky cranesbill (*G. phaeum*).
The last mentioned, although rare as a wild plant, is grown in
gardens. Its purplish-black flowers appear in May and are sometimes
covered with bees seeking nectar from morning till night.

In geraniums nectar is usually secreted at the bases of the stamens.
The pollen grain is large and easy to distinguish under the microscope, being rough with a fine network.

The true geraniums, or cranesbills, should not be confused with
the pelargoniums—to which the scarlet bedding geraniums belong—
not nearly such good bee plants.

Crocus   *Crocus spp.: Iridaceae*

Crocuses are always of value to the beekeeper producing as they do
an abundance of pollen at a time when there are few sources of fresh
pollen available. They are often grown in the vicinity of hives for
this reason. In company with the winter aconite (*Eranthis hyemalis*)
they are among the plants that are most worth while cultivating by
the beekeeper.

February is the month when the first flowers open in most districts
in the south of England, but this depends upon the forwardness of
the season and other considerations. In exceptionally early springs
crocuses at Kew have been out in the latter part of January. The
flowering period is fortunately long and extends well over a month as
a general rule.

The number of different kinds of crocus in cultivation is very large,
as a study of any seedsman's or bulb-merchant's catalogue will show.
Some of the commoner sorts are also to be found in a wild or semi-
wild state. There is probably little or no difference in their attractiveness to bees for pollen. Nectar may also be obtained by the honey
bee in some instances, but as a rule the flowers are worked only for
pollen. It has been stated (27) that some of the yellow-flowered
varieties are the best for nectar, but this requires confirmation. The

nectar is secreted at the base of the flower and as the long narrow flower-tube is almost completely filled by the style and hairs it requires a proboscis of fair length, longer than that of the honey bee, to reach it. However, if nectar accumulates and rises sufficiently in the tube the honey bee may be able to reach it by making great efforts (11).

The pollen of the crocus is generally bright orange or golden in colour. The individual grain is large, smooth and spherical and is usually easily distinguished from other pollen grains by its size, shape and colour. Some of the newer garden forms of crocus (tetraploids) have an exceptionally large pollen grain.

The so-called autumn 'crocus' (*Colchicum autumnale*), which flowers in the autumn and not in the spring, as the name indicates, also yields pollen and perhaps nectar (1). It exists wild and cultivated. The pollen is of a somewhat oily nature. See also Meadow saffron, p. 170.

Crown Imperial   *Fritillaria imperialis: Liliaceae*

This showy and stately garden plant, which grows 3 to 4 feet high and bears a cluster of large bell-like flowers at the top, is of interest on account of the large amount of nectar each individual flower is capable of yielding. The nectar is secreted in the form of six large drops at the base of the flower. Although the nectar is stated to be weak or of low sugar content (4) it is acceptable to honey bees on occasions. Flowers vary in shade from yellow to copper or red. This large bulbous plant does best undisturbed in a rich soil. It is well suited for the edges of shrubberies.

Cuckoo Flower   *Cardamine pratensis: Cruciferae*

The soft pink or pale lilac flowers of this plant are very common in meadows early in the year or just at the time when the cuckoo's song is first heard. Bees make good use of them for nectar and pollen. The nectar is secreted by two pairs of nectaries, one large and one small, at the base of the flower, and collects in the pouches formed by the bases of the sepals.

Cucumber   *Cucumis sativus: Cucurbitaceae*

The cucumber, like other members of the gourd family, is largely dependent upon bees for pollination, the male and female organs being borne on separate flowers. When grown in glass-houses, as is usual in Britain, pollination is generally done by hand. Hives of bees placed in large houses at flowering time do the work efficiently but unfortunately large numbers of bees beat against the glass continuously until exhausted and die.

When cucumbers are grown on a field scale out of doors, as is the case in some countries, crops of honey may be obtained. Such honey has been described as pale yellow or amber in colour with a rather strong cucumber-like flavour at first, which largely disappears in time (19).

Currant   *Ribes spp.: Grossulariaceae*

The black, red, and white currants of the fruit garden are all good bee plants and yield nectar and pollen fairly early in the season. The blackcurrant (*R. nigrum*) is the most extensively cultivated, and in fruit-growing districts quite large areas may be available as bee forage to beekeepers that happen to be in the vicinity. However, owing to the presence of many other nectar sources at the time flowering takes place, it is doubtful whether blackcurrant honey in anything like a pure form has ever been obtained. The enhanced value that is now attributed to blackcurrants as a source of vitamins has led to increased cultivation of this fruit.

The inconspicuous flowers of the blackcurrant have a somewhat characteristic odour. The petals are white and the tips of the sepals tinged with red. As the bell-shaped flower is only some 5 mm. deep, the honey bee has easy access to the nectar. Not only does it extract the nectar from the open flower, but may even open the older flower buds with its jaws (11).

Some of the flowering currants (*R. sanguineum*) so popular in gardens, are also good bee plants and blossom early in the season. They are generally in flower in April and seldom fail to blossom freely. Honey bees are often present in large numbers, working especially for pollen. There are many varieties of flowering currant. In some the flower-tube is too long for the hive bee to obtain nectar.

Cynoglossum   *Cynoglossum spp.: Boraginaceae*

The hound's tongue of the flower garden may include various plants (annuals, biennials or perennials) which thrive in any garden soil in a sunny position. Their deep blue forget-me-not-like flowers appear in June or July and attract bees. So also do the claret-coloured flowers of the wild hound's tongue (*C. officinale*) to be seen usually by roadsides, in waste places and around sand-dunes, etc. The nectar, protected by velvety hairs, is secreted by a fleshy nectary at the base of the flower-tube which is only some 3 mm. long (11).

Daffodil   *Narcissus spp.: Amaryllidaceae*

Daffodils, narcissi and jonquils are of some value to the beekeeper for pollen, especially as they appear so early but are of little consequence as far as nectar is concerned. The cultivated kinds vary a great deal in shape, size, colour and time of flowering. The wild daffodil (*N. pseudo-narcissus*) with its pale yellow flowers appearing usually in March, is common in moist woods and thickets in some parts of the country, especially the south-eastern counties. Nectar collects at the base of the flower-tube and is shielded or protected by the bases of the stamens and a fairly long proboscis is required to reach it. The flowers are better suited for bumble bees and *Lepidoptera* than for the honey bee.

Dahlia   *Dahlia spp.: Compositae*

Mexico is the original home of the parent plants of the many kinds of dahlia now in cultivation. Those that are single, whether tall or dwarf, large or small, are good bee plants and supply nectar and pollen at a time of the year when they are becoming short in most areas. Some of the dwarf bedding varieties now grown are particularly useful in this respect, with their long blooming period and free-flowering habit right up to the time the first frosts arrive.

Daisy   *Bellis perennis: Compositae*

The common daisy of fields and lawns that so delights the small child, and flowers most months of the year (March to October) is said to be visited by honey bees in quite large numbers for pollen in some European countries (11). The writer has never observed this to be the case in districts with which he is familiar in Britain, where the flowers appear to be neglected.

The ox-eye daisy (*Chrysanthemum leucanthemum*) with its flower-heads nearly 3 inches across and common in meadows, numbers the honey bee among its insect visitors. It is also cultivated.

Daphne   *Daphne mezereum: Thymelaeceae*

The common daphne or mezereon is a shrub often seen in gardens, especially cottage gardens, and is favoured for its sweet-scented flowers that appear so early in the year and before the leaves. It is usually in flower early in February and so keeps company with the crocus and snowdrop. A white-flowered variety is also grown. The mezereon exists apparently wild in woods in the south of England, but is actually a native of Europe and Siberia. Honey bees frequent the flowers when the weather is warm enough for nectar and pollen.

The flowers of another daphne, the spurge laurel (*D. laureola*), an evergreen shrub usually to be seen in copses and woods on chalk or limestone, also attract bees.

Date Plum   *Diospyros lotus: Ebenaceae*

This name is used for a small deciduous tree sometimes seen in cultivation, but generally only in collections. It has a wide distribution in the wild state, extending from North Africa and Asia Minor to China. The fruit is edible. Small unisexual flowers appear in June and are actively worked by honey bees for nectar. At least this is the case year after year in the Kew area. The tree is quite hardy and although introduced to cultivation in England in the seventeenth century has never been much planted.

Other species of *Diospyros* are known to be good nectar sources for the honey bee, including the American persimmon (*D. virginiana*) which is also seldom seen in cultivation in Britain.

Deadly Nightshade (formerly **Belladonna**)
                          Atropa bella-donna: Solanaceae

This important drug plant which is sometimes, although not frequently, to be found in the wild state, is not generally regarded as a bee plant. However, belladonna plots at Kew during the World War II were much visited by honey bees when in flower in May, obviously for nectar. The bee climbs right into the large bell-shaped purple flower, out of sight, and is easily able to reach the nectar which is secreted by a conspicuous annular yellow nectary at the base of the ovary. It has to thrust its proboscis between the hairy lower part of the stamens to get at the nectar. In crawling in and out of the flower it becombes dusted with the pale, cream-coloured pollen.

Deadnettle   *Lamium spp.: Labiatae*

Honey bees are sometimes to be seen around the flowers of dead-nettles, but the nectar is too deep-seated for them, only bumble bees being able to reach it. When punctures exist at the base of the flower, as sometimes happens, they may be able to obtain a certain amount. The flowers may also be visited for pollen, especially those of the red deadnettle (*L. purpureum*) so common everywhere in meadows, along roadsides and sometimes in masses on hedge banks. This plant begins to flower as early as February when few pollen plants are available and remains in flower throughout the summer. The pollen is a beautiful dark orange in colour.

Discaria   *Discaria serratifolia: Rhamnaceae*

This deciduous shrub from Chile with long, excessively spiny, pendulous branches is not often seen in cultivation, although quite hardy. It bears clusters of small greenish-white flowers in June or July which are sweet scented and attract honey bees in large numbers.

Dodder   *Cuscata spp.: Convolvulaceae*

These twining parasitic plants are often to be seen attached to wild plants, such as gorse and heather. They may also prove troublesome pests with cultivated plants, especially clover and flax. There is evidence that the small fleshy flowers which appear from June onwards provide the honey bee with nectar (23).

Dogwood   *Cornus spp.: Cornaceae*

Some of the many dogwoods in cultivation at Kew attract bees, especially the Siberian dogwood (*C. alba* 'Sibirica') sometimes grown in gardens. The cornelian cherry which is covered with yellow blossoms in February and March, does not appear to be very popular with bees in spite of its early flowering. It may be visited more in some European countries (1).

Doronicum   *Doronicum spp.: Compositae*

Bees visit the sunflower-like flowers for nectar and pollen. The most useful to the beekeeper are undoubtedly those that flower early in the spring, from March onwards (such as *D. plantagineum excelsum*) which are in fact the most generally cultivated. The doronicums are of vigorous growth, thrive in any soil, and are well suited for rough places and for naturalizing. Two species do in fact occur apparently wild in some parts of the country.

Dorycnium   *Dorycnium rectum: Leguminosae*

The pink flowers of this small shrubby plant from southern Europe
are freely worked for nectar at Kew in July. It does not exceed 2 feet
in height, grows easily from seed and, like so many Mediterranean
plants, succeeds on a light dry soil.

Dracocephalum   *Dracocephalum moldavica: Labiatae*

This is a small annual with blue and white flowers, resembling a
salvia and belonging to the same family. It is a native of eastern
Siberia, but grows well in Britain. The crushed leaves are aromatic
and have been used in the same way as those of bee balm (*Melissa*)
for rubbing inside skeps and hives to attract swarms (*Bee World*,
1931, 80).

Echinops   *Echinops spp.: Compositae*

The globe thistles or echinops with their large prickly heads of blue
flowers never fail to attract bees in the summer. The name echinops
is from Greek meaning 'like a hedgehog', which is of course appropriate. There are eighty or more species of the genus and they occur in
the wild state mainly from Spain and Portugal eastwards to India.
Honey bees have been observed visiting the flowers for nectar, in
company with numerous bumble bees and wasps, in the following:
E. bannaticus (common in gardens); *E. exaltatus* (Hungary, flowers
white); *E. niveus* (Himalaya); *E. ritro* (Mediterranean region); and
E. sphaerocephalus (south Europe). These thistle-like plants are well
suited for herbaceous borders and shrubberies and are all of easy
culture.

   E. sphaerocephalus has long been grown in gardens and garden
forms or varieties such as *giganteus* and *albidus* exist. Many years ago
this species attracted a good deal of attention among beekeepers,
especially in the United States, where a Mr H. Chapman of Versailles,
New York, planted about 3 acres of it. Glowing reports were
circulated about the plant, which is undoubtedly very attractive to
bees, and from that time it became known as the Chapman honey
plant (19). This name is used by British beekeepers interested in bee
plants.

   Although the plant attracts bees so much, some observers doubt
whether it is really of great value as a nectar plant (9), the well-known
American beekeeper Dr C. C. Miller stating: 'I never saw bees so

thick on any other plant. But close observation showed that the bees were not in eager haste in their usual way when getting a big yield, but were in large part idle. It looked a little as if the plant had some kind of stupefying effect on them' (*Gleanings in Bee Culture*, Dec., 1918).

It has been suggested that the Chapman honey plant makes a good apiary hedge or boundary plant if given two wires for support, the dead stalks being left in position through the winter.

In the individual flowers of this plant the nectar is secreted at the base and may rise in the corolla tube, which is 5 to 6 mm. long and largely occupied by the style, and overflow into the bell or expanded part of the flower. As this is split almost to its base into the corolla lobes or petals the nectar is easily available to honey bees.

Elder   *Sambucus nigra: Caprifoliaceae*

The strongly-scented flowers of the elder, which most people find unpleasant, are not normally visited by the honey bee and have been described as nectarless (11). The fact that they appear in June when there is usually an abundance of other pollen sources may be the reason why they do not even attract the honey bee for pollen. As the plant is so common everywhere, even to the extent of becoming a nuisance and a pest in some districts, and flowers so freely, it is a great pity for the beekeeper that it is not a good honey plant.

Elm   *Ulmus spp.: Ulmaceae*

Elms are sometimes useful early sources of pollen, particularly when they happen to be growing in close proximity to an apiary. The purple clusters of small flowers appear in February or March and may pass unnoticed as they are so often on tall trees. Pollen is produced in abundance and is wind-borne. In favourable weather at this early season bees will often make good use of it.

In California the introduced European elms are a valued source of pollen to beekeepers, the trees being alive with bees (23) in spring, when better bee-flying weather is likely to prevail than in Britain.

There seems to be some evidence that in warmer climates elm flowers may also be a source of nectar (23; 4). Later in the year the elm sometimes becomes a source of honeydew.

Elsholtzia   *Elsholtzia spp.: Labiatae*

Some good bee plants are to be found in this genus which belongs to the mint family and should not be confused with *Eschscholtzia* (Californian poppies) which belong to an entirely different family. The most interesting is perhaps a Chinese shrub (*E. stauntonii*) which was only introduced to cultivation in Britain in 1909, and which has acquired the name of heathermint owing to its purple (heather-like) flowers and aromatic (mint-like) leaves. Much has appeared in regard to it in the American bee press.

The shrub reaches 5 feet in height, needs full sunshine and grows in almost any soil except heavy clay. It is hardy in the south of England, the shoots often dying back in the winter, but fresh ones arising in the spring. It flowers freely in October or the end of September, producing clusters of fragrant flowers 3 to 8 inches long, at the ends of the branches. In the individual flowers the funnel-shaped corolla tube is only about a quarter of an inch in length and well suited to honey bees who work it freely for nectar, as do bumble bees. Much depends, of course, on the nature of the weather so late in the year. It is undoubtedly a useful and ornamental late-flowering shrub for bees and well worth a place in the bee garden.

E. crispa, also from the Orient, is an annual 1 to 1½ feet high and better known in cultivation. It is also a good nectar plant (*American Bee Journal*, Nov., 1934).

E. cristata, which is an annual, is perhaps better known in cultivation and is another good nectar plant. It grows 1 to 1½ feet high and produces numerous lavender-coloured flowers on one-sided spikes in the late summer, which bees eagerly visit. The plant has aromatic leaves and is easily grown, requiring full sun.

Endive   *Cichorium endivia: Compositae*

When this salad plant is grown for seed or left neglected in the vegetable garden the blue flowers, which appear at intervals along the stem, are visited by bees for nectar. The flowers are not unlike those of chicory, a close relative of the plant.

Erigeron   *Erigeron spp.: Compositae*

These popular border perennials generally produce their flowers in July. They are pink or purple and rather like an aster. Among the better known of these fleabanes, *E. speciosus*, *E. macranthus* and *E. multiradiatus* have been observed to be freely visited by honey bees for nectar, as are doubtless many others. The first of these is perhaps the best of the taller kinds, reaching 2½ feet with masses of large purple flowers.

Erodium   *Erodium spp.: Geraniaceae*

Known also as stork's bill or heron's bill, these plants are good nectar yielders, as are the hardy geraniums, their near relatives. Three species occur wild in Britain, generally most common near the sea, and others are cultivated in gardens. One of the wild species (*E. cicutarium*) is very prevalent in parts of Europe and has become widely naturalized in North America, where it is known as pin clover or filaree, and 'produces an abundance of pollen and considerable honey of good quality' (19). Another British species freely naturalized in other countries which has proved a good pollen and nectar plant is the musky stork's bill (*E. moschatum*) (23).

Escallonia   *Escallonia spp.: Escalloniaceae*

The escallonias are a group of handsome shrubs, natives of South America. They are mostly evergreen and several hail from Chile and Peru. Unfortunately the majority are too tender for most parts of Britain, except in the mildest parts or with the protection of a wall. A number are good bee plants. They are more commonly cultivated in the south-western counties, especially *E. rubra macrantha*, which is often used for hedges near the sea.

E. 'Langleyensis' (a hybrid) is fairly hardy and grows well at Kew, where its pretty rosy-red flowers are a constant source of delight to honey bees. The flowers appear in profusion in June and July, while a few may still be found as late as September. It is bushy, reaching 8 feet or more in height with slender arching branches—altogether a handsome shrub. Nectar is secreted freely at the base of the flower-tube.

Eucryphia   *Eucryphia glutinosa: Eucryphiaceae*

This evergreen or partly evergreen shrub from Chile bears large solitary flowers in July and August. These are striking, with their pure white petals and masses of stamens. Bees visit them for pollen and for nectar. The shrub can hardly be described as common in cultivation, probably because it is not among the easiest of plants to propagate and transplants badly.

Eupatorium   *Eupatorium spp.: Compositae*

A number of rather coarse garden perennials belong to this genus, some better suited to the wild garden than the flower border. For the most part they flower in the autumn. Bees work the flowers of a number of species freely for nectar and pollen. Some of those that originated in North America are regarded as good bee plants there (19).

Honey bees also visit the small flesh-coloured flowers of the solitary wild British species of *Eupatorium*, i.e. hemp agrimony (*E. cannabinum*), a very common plant along streams and in moist places generally. The flowers appear in dense terminal heads in July and August and are popular with butterflies, especially painted ladies and red admirals.

Evening Primrose   *Oenothera biennis: Onagraceae*

This stately plant, in one or other of its many forms, is to be seen both in the flower garden and as a weed in waste places. It is a naturalized and not an indigenous plant. The large, attractive yellow flowers open in the evening. They are normally pollinated by night-flying moths but bees often visit them, especially in the early morning in the summer for pollen.

The pollen grains of the evening primrose are very large and easily visible to the naked eye. They are bound together by yellow threads (strands of viscin) which may sometimes be seen trailing on the legs of bees working the flowers. According to one observer (27) bees will sometimes settle on the flowers and drink from the drops of water that have collected there after a shower of rain. There are many varieties of evening primrose in cultivation.

Eyebright   *Euphrasia officinalis: Scrophulariaceae*

The brightly-coloured flowers of this variable little plant may receive the attention of the honey bee. It is common in all kinds of situations, including dry meadows and exposed hillsides or cliffs near the sea. Nectar is secreted by the lower part of the ovary and collects in the corolla tube.

False Indigo   *Amorpha fruticosa: Leguminosae*

Honey bees make use of the purplish-blue flowers of this shrub for nectar and pollen. It is not often cultivated, being of a somewhat clumsy, straggling nature. It reaches 10 to 15 feet in height, but is generally cut back to ground level by frost in winter in English gardens. The flowers appear in July in clusters and the foliage is ornamental. A native of the southern United States, this shrub is naturalized in some parts of Europe.

Fennel   *Foeniculum vulgare: Umbelliferae*

The flowers of this herb attract bees and there are records of honey being obtained from it in other countries (23). It is often cultivated in gardens and smallholdings, the leaves being mainly used for flavouring fish sauces, especially for boiled salmon and mackerel.

Like most plants of this family the heads of small white flowers swarm with flies and other short-tongued insects who are able to get at the relatively exposed nectar but honey bees are not such frequent visitors as they are with many other plants. This was plainly visible in large plots of fennel grown for seed at Kew during the World War II, where flowering took place in June and July.

The flowers of the wild fennel are similar to those of the cultivated plant. It is common in many localities, especially near the sea, where its bluish-green stems and finely divided hair-like leaves make it a conspicuous plant.

Figwort   *Scrophularia spp.: Scrophulariaceae*

Some of the wild figworts are first-class bee plants, their flowers secreting nectar very freely. The best known in this respect is the knotted figwort (*S. nodosa*), so named on account of its knotted or tuberous roots. It is a coarse perennial plant, 3 to 4 feet high, found in woods and moist places, and bearing much-forked bunches of small globular flowers. These can hardly be termed attractive, being dull purple in colour and tinged with greenish yellow. The nectar is secreted in large drops at the base of the corolla by a circular swelling, and is protected from rain. Sometimes it collects to such an extent as to half fill the flower (27). Besides honey bees, wasps are very frequent visitors to the flowers. The water figwort (*S. aquatica*) which is very similar and grows by ditches and streams, is also a good nectar and pollen plant for bees. The yellow figwort (*S. vernalis*) is a smaller plant with more ornamental flowers. It is in bloom in spring but is rare in most districts.

An American figwort (*S. marylandica*) closely allied to *S. nodosa*, has also a high reputation as a bee plant and as a profuse nectar yielder. Under the name of Simpson's honey plant it attracted a great deal of attention at one time and was considered to be one of the best plants for artificial pasturage for bees. Honey bees were found to visit the flowers in a small field of the plant in great numbers from morning till night. On an average each flower was visited at times at the rate of one a minute, and after the nectar was removed other drops would be secreted in about two minutes (20).

Flax   *Linum usitatissimum: Linaceae*

The flax plant is cultivated in most temperate countries both for its seed, which yields linseed oil and linseed cake for stock feeding, and for flax fibre, the raw material of the linen industry. It has been grown in Britain for many centuries, but linseed has never been the important crop that it is in other countries. The plant is an annual with an attractive blue flower. A field of linseed in full bloom is indeed a pretty sight. Bees visit the flowers, but it is doubtful whether they are ever important sources of nectar, although some observers report having seen the flowers visited very freely. One stated a seven-acre field of linseed had been a great attraction 'for the bees' morning shift' but that most of the petals had fallen by the afternoon (*Gardening Illustrated*, 29th August 1941). The pollen grain of flax is a large spherical one with a cellular covering (27).

The wild and the cultivated flax plants of the flower garden appear to be of about the same value as bee plants. Purging flax (*L. catharticum*) is a very common plant in meadows and on chalk hills and cliffs. It is in bloom from June to August and has small white, not blue, flowers. In these and other Linums the stamens are fused together at the base into a fleshy ring where the nectar is secreted in five drops. It is produced from five small pits or nectaries.

Fleabane   *Pulicaria dysenterica: Compositae*

The yellow flower-heads of this wild plant, rather like a small marigold, are visited by honey bees to some extent. They are available for a long period in most areas—from July to September—and are usually to be found near streams and in moist places.

Forget-me-not   *Myosotis spp.: Boraginaceae*

The forget-me-nots, whether wild or cultivated, are always popular with bees. Among the wild kinds the field forget-me-not (*M. arvensis*) is the most abundant, its small saphhire flowers being common in fields and woods alike. A larger and more attractive flower is that of *M. scorpioides*, the true forget-me-not, and the plant which the legends refer to. It is also wild in many parts of Europe and known by the same name (e.g. 'vergissmeinnicht'—German). Its dainty blue flowers are most prevalent on the banks of rivers and streams.

Where forget-me-nots abound the pollen of the plants is often very prevalent in honey. The reason given for this is that the mouth of the flower tube is very narrow and there is barely room for the bee to insert its proboscis. In doing so it is forced to dislodge a good deal of pollen which must get mixed with the nectar. The pollen grains of forget-me-not are exceedingly minute, among the smallest known in fact, and so are probably drawn up with the nectar. They also get lodged in the hairs of the proboscis. The net result is that they reach the honey stomach in much greater quantity than is the case with the pollen of other flowers (*Bee World*, 1943, 23).

The pollen grains of forget-me-not may measure no more than three to four microns in length and compared under the microscope with large pollens such as crocus, mallow, hollyhock, etc., look rather like a marble next to a football. In shape they are not round, but appear as two small spheres joined together by a narrow neck.

Fraxinella   *Dictamnus albus: Rutaceae*

This European plant, a favourite in the flower garden in bygone days, is less grown now than formerly. It is also called candleplant and burning bush, because the oil secreted by the plant on hot calm days is inflammable. Bees visit the pale purple (or white) flowers for nectar and pollen.

French Honeysuckle   *Hedysarum spp.: Leguminosae*

The flowers of several kinds are visited by bees for nectar. Two of the best known in gardens are *H. coronarium* with red or white flowers, and *H. obscurum*, a much smaller plant with racemes of showy purple flowers.

Fuchsia   *Fuchsia magellanica: Onagraceae*

This South American plant is hardy in the milder parts of Great Britain and attains the dimensions of a small tree if unpruned. In less mild districts it may be killed to the ground in the winter but sends up fresh-flowering shoots each year. In Devon and Cornwall and western Ireland it grows vigorously and has become naturalized in some areas. It is quite commonly used there as a hedge plant, even for farm hedges, for it makes bushy compact growth and withstands the sea breezes. The red and purple flowers are rich in nectar which the honey bee is not slow to utilize. They are said to be an important source of nectar in Ireland. The pendent nature of the flower protects the nectar well from rain. At Kew, beds of the plant attract bees in large numbers in August year after year. Honey from fuchsia is described as very light in colour but of little flavour (*American Bee Journal*, 1941, 252).

Gaillardia   *Gaillardia spp.: Compositae*

Available in annual or perennial forms, this very popular garden plant is attractive to bees. They are often to be seen probing the flower-heads for nectar, or collecting pollen. Some varieties have the advantage of flowering more or less continuously throughout the summer. *G. pulchella* is one of the main honey-producing plants of Texas (20).

Galtonia   *Galtonia candicans: Liliaceae*

The white bell-like flowers of this Cape plant are reputed to attract bees freely (11). Known also as Cape hyacinth, this tall, bulbous plant is sometimes to be seen in English gardens. It is hardy in light soils and of easy culture. Flowering takes place in August and September, the nectar collecting at the base of the flower with six fairly deep nectar passages leading to it.

Garrya   *Garrya elliptica: Garryaceae*

The male catkins of this evergreen shrub are sometimes visited by bees for pollen. It grows best in the warmer districts—Devon and Cornwall—where the catkins reach a foot in length.

Gaultheria   *Gaultheria shallon: Ericaceae*

This evergreen shrub of the heath family prefers moist, shady spots and will grow in any ordinary garden soil, but does best in one of a peaty nature. It forms a dense low thicket and spreads by underground suckers. On account of its density it has been cultivated for game cover in some parts of the country and so exists apparently wild. The pinkish-white flowers appear in May and June in small racemes or bunches. They are a good source of nectar and in its native land, western North America, the plant is regarded as one of the best honey plants, being commonly called shallon.

Geum   *Geum spp.: Rosaceae*

The geums of the rock garden and flower border are visited by bees, mainly for pollen. The same applies to the wild species. In the water avens (*G. rivale*) large drops of nectar may be secreted on the receptacle of the flower, which honey bees procure from the outside (11).

Gilia   *Gilia spp.: Polemoniaceae*

These hardy annuals with their colourful flowers of various shades are in bloom for a long time, and also last well in vases. In the milder districts they may be sown in the autumn for early blooming the following season, and in the bee garden are perhaps best grown in this way. In parts of North America, their original home, the wild gilias are useful sources of nectar, especially in burnt-over forest country, and yield surplus honey (23). Many gilias are regarded as good nectar plants. Some yield pollen of a blue colour. Certain species are better sown in spring or Summer.

Gipsywort   *Lycopus europaeus: Labiatae*

Although not common in all parts of the country, this plant is prevalent in some areas, particularly along the margins of rivers and pools. The small white flowers are in clusters on the upper part of the stem and are present from June to September. They are rich in nectar, which the honey bee has no difficulty in extracting, and are a useful late though minor source of nectar in some districts. Deeply-cut leaves characterize the plant and the name gipsywort arose from the fact that gipsies used the root to stain their faces brown.

Glory of the snow   *Chionodoxa luciliae: Liliaceae*

This little plant, which comes from Asia Minor, is one of the most handsome of the early spring-flowering bulbs and is quite hardy. Its blue and white or pure white flowers appear in March and April and are useful for early pollen and for nectar if plentiful. Any ordinary garden soil will suit it and it increases rapidly.

Goat's Rue   *Galega officinalis: Leguminosae*

This perennial from southern Europe is often to be seen in gardens and has been listed as a bee plant (27). However, the writer has never observed honey bees visiting the pea-like flowers even on fine sunny days when bees have been flying freely. The pollen grains are very small, not much larger than those of forget-me-not (*Myosotis*) (27). Bumble bees sometimes visit the flowers for pollen.

Godetia   *Godetia spp.: Oenotheraceae*

In some instances the flowers of these everyday garden plants supply the hive bee with nectar and pollen. They vary much in size, shape and colour, while some have single flowers and others double.

Golden Honey Plant   *Actinomeris squarrosa: Compositae*

On the North American continent this perennial plant is widely distributed and has been favourably reported on as a honey plant (*Canadian Bee Journal*, May, 1937). It grows best there on rich lowlands, reaching a height of 5 to 8 feet and blooming in August and September. Its yellow flowers are freely worked by honey bees. As it is somewhat coarse in appearance it does not appear to have been grown in gardens in Britain, but is an interesting possibility for the bee garden.

Golden Rod   *Solidago spp.: Compositae*

There are about a hundred different species of *Solidago* or golden rod, the great majority of which are natives of North America, where they are important autumn honey plants. One species, how-ever, *S. virgaurea*, is wild in Britain and is prevalent in some areas, particularly near heaths and on rocky banks and cliffs near the sea. The same species occurs freely in many European countries, especially southern Norway, where it is sometimes a dominant plant over large tracts of country. It grows in the poorest soils and produces its abundance of yellow flowers from July to October. These are very attractive to honey bees as well as to numerous other insects and must constitute a useful source of late nectar where they are suffic-iently abundant.

The same applies to the many garden forms of golden rod of American origin that are so popular in herbaceous borders and the shrubbery or wild garden where they are well able to hold their own with other plants. Some of these introduced kinds have become naturalized to some extent in Britain. The following golden rods, all North American, have been observed at Kew to be freely worked for nectar and pollen by honey bees late in the season: *S. elliptica, S. graminifolia, S. latifolia, S. pilosa, S. puberula, S. riddellii, S. rigida, S. shortii.*

In many parts of eastern Canada and the United States golden rod and aster are the two important late-season honey plants. Their main value is in providing winter stores for bees, but surplus honey is taken from golden rod in some areas. This is described as golden yellow in colour, thick and heavy and of fine flavour. The flavour improves markedly during ripening. It crystallizes in about two months with a coarse grain (20). Little is known regarding the nectar value of the actual individual species, but some contend that all golden rods will yield nectar under conditions favourable to them. However, in some parts of the United States where golden rod is common it yields little or no honey. Some contend that a fairly high temperature is necessary for a good nectar flow with golden rod (*American Bee Journal*, 1936, 592).

Gooseberry   *Ribes grossularia: Grossulariaceae*

The gooseberry occurs as a wild plant in the north of England. It is the progenitor or parent plant of the numerous varieties now cultivated with rough or smooth, green, yellow, red or purplish berries. The gooseberry has been cultivated in Britain for many centuries, every cottage garden, however small, having its own gooseberry bushes at one time. These sometimes attain a large size and great age, fifty years or more.

The leaves of the gooseberry are among the first to show signs of growth in early spring, and the greenish-white, inconspicuous flowers are generally out in April. These are attractive to honey bees for their nectar, which is secreted at the base of the bell-shaped flower, and is protected by the hairs projecting from the style. The hairs may impede other insects but offer no obstruction to the hive bee. The flower is in fact well constructed for its visits.

Surplus honey from gooseberries may be obtained from hives near fruit farms where gooseberries are extensively grown, but this probably rarely occurs. However, such honey has been described as medium coloured and of excellent flavour (27). It has been found that where currants and gooseberries are interplanted with cherries, as is sometimes done, bees may desert the currant and gooseberry flowers in favour of those of the cherries (*Royal Horticultural Society Journal*, LI, 231). The gooseberry is regarded as a better nectar plant than the currant. The pollen of the gooseberry is not produced in great abundance and is pale greenish yellow in colour. It is not in separate granules like that of the common tree fruits, but is stuck together in glutinous masses. That of the currant is similar. It is collected as a side line while bees work for nectar.

Gorse   *Ulex europaeus: Leguminosae*

The prickly gorse or furze bush with its familiar yellow flowers is one of the commonest of British plants and covers many thousands of acres of moors, commons and heathland, particularly in the south of England. It thrives on poor, light sandy soils, often coming up freely wherever soil is disturbed and is to be found in flower throughout the year as a rule. However, spring (April) is the main flowering period, when the bushes are covered with their golden blossoms.

Gorse flowers are undoubtedly a useful standby for pollen for beekeepers in many parts of the country. Whether they supply nectar as well is rather open to question, but the opinion of careful observers is that they probably do at certain times. However, it is for pollen that the plant is mainly of value. This is produced in abundance and is bright yellow or orange in colour, assuming a darker or duller shade in the bees' pollen baskets (8). In the early spring it is especially useful. Flowers often commence to appear freely as early as February, and so may take the place of willow and hazel in other districts. Bees commonly forsake gorse once other flowers become available.

The dwarf gorse (*Ulex minor*, formerly *U. nonus*) is also very common on heaths and commons in the southern counties. It differs from ordinary gorse in its more trailing habit, smaller size and smaller flowers, the latter being about half the size of ordinary gorse and probably more easily worked by the honey bee. It flowers from June to December, generally freely in September when those of ordinary gorse are mainly over. In company with heather or ling (*Calluna*) it is the main source of pollen in August and September in some heathland areas (e.g. Bagshot, Surrey) the two pollens being often found mixed in bees' pollen baskets (Miss Betts, *Bee World*, Oct., 1935).

Gourds   *Cucurbitaceae*

All the gourd plants are useful sources of nectar and pollen to the honey bee. They include vegetable marrows, pumpkins, cucumbers, squashes, melons and the ornamental gourds sometimes grown in gardens for their decorative value or quaint appearance. Only the first of these can be said to be commonly or generally cultivated out of doors in Britain (see vegetable marrow, cucumber, etc.).

Grape   *Vitis vinifera: Vitaceae*

During the Middle Ages there were vineyards in the milder parts of Britain, mainly at monasteries, but from the late Middle Ages to the end of World War II the grape was seldom grown seriously in the open. However, during the last twenty to thirty years, outdoor cultivation of vines has been successfully resumed in southern and western Britain.

The flowers of the grape are small and inconspicuous, but their strong fragrance, suggestive of mignonette, compensates for this. The structure of the flower is interesting in that the five petals remain united at the top and become detached at the base, falling away as a cap. The stamens produce very little pollen. Alternating with them are five nectaries. These secrete nectar in warm, but apparently not in cold, climates. In grape-growing countries honey bees may collect nectar and pollen from the vineyards. When the grapes are ripe the juice of damaged fruit may also be freely collected when nectar is scarce. This sometimes ferments in the hive (23).

Grape Hyacinth   *Muscari spp.: Liliaceae*

The pretty blue flowers of these ever-popular spring bulbs are very attractive to bees early in the year and yield pollen and nectar, the nectar being secreted in the septal glands of the ovary. There are many different kinds of grape hyacinth in cultivation. All are of the easiest culture and hardy, which is rather surprising, seeing that they are for the most part of Mediterranean origin. One species (*M. racemosum*) occurs wild in some parts of the country but is not common.

Grasses   *Gramineae*

Grasses are essentially wind-pollinated plants, without nectar, and yield a light powdery type of pollen. Bees do not seem to favour this type of pollen as a rule when others are available. However, they do sometimes collect pollen from grasses. There are many well-authenticated cases of this. Probably any grass-producing pollen may be visited freely, but the two grasses often concerned are cock's foot (*Dactylis glomerata*) and meadow foxtail (*Alopecurus pratensis*). The last mentioned has been observed to be worked for pollen at Kew in May, a time when numerous other pollen sources are available in the area, which goes to prove there is no accounting for the ways of bees.

Greenweed   *Genista tinctoria: Leguminosae*

The yellow gorse-like flowers of this plant attract the honey bee for pollen at times, but do not appear to yield nectar. It is often common on clay soils. The young tops were at one time used for dying wool, hence the name dyer's greenweed by which it is known. An allied plant petty whin (*G. anglica*) that occurs on moors and heaths is also worked for pollen.

Grindelia   *Grindelia spp.: Compositae*

This group of North American plants, with yellow flower-heads 1 to 2 inches across, have been grown in English gardens but are not generally cultivated. The most interesting from the beekeeper's point of view is perhaps the gum or rosin weed (*G. squarrosa*), common in the southern USA and a source of honey of somewhat inferior flavour and granulating very quickly (23). This plant when grown at Kew is much sought after for nectar. The flower-heads are covered with a peculiar milky gum.

Gromwell   *Lithospermum spp.: Boraginaceae*

Both the wild and garden species of gromwell or borage-worts (mainly rock garden plants with blue flowers) are sometimes visited by hive bees, and often flower in great profusion. Nectar is secreted by the ovary and collects at the base of the corolla tube, which in the case of the corn gromwell (*L. arvense*) is $4\frac{1}{2}$ mm. long. They are probably nowhere of special importance as honey plants. The corn gromwell, bearing white flowers in May and June, occurs as a weed in cultivated ground.

Ground Ivy   *Glechoma hederacea: Labiatae*

This plant, which is common in so many different situations, and often to be seen trailing in masses from old walls and on dry banks, may be in flower very early in the year, sometimes towards the end of February. Honey bees visit the flowers for nectar and pollen (27) but these are not borne in great profusion and are probably of little importance.

Groundsel   *Senecio vulgaris: Compositae*

This is perhaps the commonest weed of cultivation, and few gardens are without it. It may be in bloom early and late in the year. Some writers (27) state hive bees visit the flowers, but they do not seem to offer much attraction, and at most seasons of the year are entirely neglected.

Gypsophila   *Gypsophila paniculata: Caryophyllaceae*

The myriads of tiny white flowers of this well-known perennial are visited by honey bees for nectar in July and August. This applies only to the single form and not to those with double flowers. The flowers of *G. repens*, a small perennial of the rock garden, also attract bees, as may those of other gypsophilas.

Hawk's Beard   *Crepis spp.: Compositae*

These plants, with flowers like those of a small dandelion, are common weeds of cultivated ground. Hive bees visit the flowers for nectar and pollen in July and August but they can only be regarded as very minor sources in most areas.

Hawk's Bit   *Leontodon autumnalis: Compositae*

Like the hawk's beard the yellow flowers of this plant are also visited by bees throughout the season.

Hawkweed   *Hieracium spp.: Compositae*

Honey bees have been observed collecting nectar and pollen from some of the many different kinds of hawkweed (especially *Pilosello officinarum*—formerly *H. pilosella*—and *H. umbellatum*) that are so widely distributed throughout the country.

Hayrattle   *Rhinanthus minor: Scrophulariaceae*

The corolla or flower-tubes of this meadow plant are, like those of its near relative the red bartsia, too long for the honey bee, but are frequently punctured near the base and the honey bee takes advantage of this. Hayrattle is also known as yellow rattle and corn rattle.

Hazel   *Corylus avellana: Corylaceae*

Like the hawthorn the hazel is very widespread throughout the country but avoids acid peaty soils, and even extends to the extreme north of Scotland. In oak and ash woods it is common, being often grown for coppicing, and flourishing best on calcareous soils where pure thickets or copses of it may occur.

It is as an early source of pollen that the hazel interests the bee-keeper. The tassels of male flowers are conspicuous from late January until early March according to season and district. They yield an abundance of a light-yellow powdery pollen, that bees will collect eagerly in suitable weather when hazel bushes are near their hives. It is unusual for them to visit bushes a long way from home at this early season. The flowers are wind pollinated and not dependent on insect visits for fertilization. They are available for quite a long period, usually about a month, but this is governed by weather conditions at flowering time.

In some parts of the country, notably Kent, hazel-nut and cob-nut orchards exist. The varieties grown in these orchards are fundamentally the same with regard to flowering as the wild hazel bushes except that the tassels are sometimes much longer.

Helenium   *Helenium spp.: Compositae*

The yellow or bronze flowers of this perennial attract bees in large numbers for nectar and pollen. Their value in the herbaceous border is mainly for autumn flowering, when they keep company with golden rod and Michaelmas daisies. Some of the newer dwarf kinds such as 'Crimson Beauty' and 'Moerheim Beauty' flower earlier and for a longer period. A bed of these plants will swarm with bees all day in good weather.

The value of the helenium to the beekeeper in Britain, however, can only be the same as that of other autumn garden flowers—in helping to build up stores for the winter. It is interesting to note that in their native land, North America, some heleniums have a reputation for yielding bitter honey, a small quantity of which will spoil other honey, although as a winter food for bees it is perfectly wholesome.

Heliotrope   *Heliotropium spp.: Boraginaceae*

The delicate fragrance always associated with the garden heliotrope is well known and it is not surprising that honey bees should think the flowers worth a visit. Some of the varieties used for bedding at Kew are a great attraction to bees for nectar in August.

Hellebore   *Helleborus spp.: Ranunculaceae*

These hardy perennials, which bloom in the winter or early spring when there is little else in flower, are attractive to bees on those occasions when the weather is sufficiently mild for them to fly. They fall into three groups—white, green and red flowered.

The old-fashioned white-flowered Christmas rose (*H. niger*), is the first to flower and may be in bloom at Christmas. Other kinds flower in sequence finishing with the red-flowering sorts in the spring.

The two wild British species, *H. viridis* (the green hellebore) and *H. foetidus*, although hardly garden plants, are useful for the wild garden, flowering in March. Both yield nectar freely (27). The flowers of all hellebores are interesting in that the petals are modified into raised funnel-shaped nectaries, which hold the nectar perfectly, like miniature vases. What appear to be the petals of the flower are actually the coloured sepals. This same ingenious type of nectary is also to be seen in the winter aconite, another early spring-flowering plant of the same family.

Hemp Nettle   *Galeopsis tetrahit: Labiatae*

Bees visit the flowers of the hemp nettles, but only for pollen, the nectar being too deep-seated. The common hemp nettle with its purple flowers is sometimes prevalent in the stubble of cornfields and is in bloom from July onwards.

Henbane   *Hyoscyamus niger: Solanaceae*

This important drug plant does occur wild or semi-wild in Britain but is not common, nor is it generally regarded as a bee plant. It may be grown on a field scale for medicinal purposes. When this was done during the World War II honey was said to have been stored from it in one instance. The field in question was cut while in flower and several workers got stung in the process. Bees kept returning to the bare patch where the plants had recently been. The honey alleged to have been obtained from it was of quite good flavour but crystallized very soon.

Heracleum   *Heracleum spp.: Umbelliferae*

These coarse-growing perennials usually attract attention on account
of their large size and commanding appearance but are better suited
for odd corners rather than the flower border. Bees visit the flowers
and compete with numerous other insects for the nectar. The plants
can only be regarded as second-rate bee plants, although sometimes
recommended. *H. villosum* (formerly *H. giganteum*) and *H. man-
tegazzianum*, a giant species from the Caucasus, 10 feet high with
flower-heads a yard across, are among those that have attracted
attention in this respect (*Bee World*, 1925, 154).

The common hogweed or cow parsnip (*H. sphondylium*) of the
hedges and meadows is visited occasionally for pollen.

Heuchera   *Heuchera spp.: Saxifragaceae*

These tufted perennials are usually grown for the sake of the leaves
rather than for their flowers. Bees have been observed visiting the
flowers for nectar.

Hibiscus   *Hibiscus syriacus: Malvaceae*

This shrub exists in numerous cultivated varieties but the species
is the only one that is hardy in the climate of Britain. Flowers may be
white or any shade of red, blue, purple or striped, and vary in size up
to 4 inches across. They do not appear until late in the season but
may be visited for pollen.

Holly   *Ilex aquifolium: Aquifoliaceae*

The holly is the commonest evergreen tree in the United Kingdom
and is to be found throughout, except in the extreme north. It is
abundant in woods, especially in the West Country, where the wet
climate seems to suit it. It more often occurs as a shrub than a tree,
although trees with large trunks do occur. The holly is also found
wild in most countries of Europe.

Flowering takes place in May or June but the small fragrant white
flowers, in clusters in the leaf axils, are by no means conspicuous and
would pass unnoticed by some people. They secrete nectar freely and
this is much sought after by the honey bee to whom it is easily
accessible. Holly is probably a much more useful nectar source than
many English beekeepers realize, especially as it may be in flower
between fruit blossom and clover. It is unfortunate that the flowering

period is rather short, generally two to three weeks, and that the trees or bushes do not always flower freely. The holly is strange in this respect for sometimes a tree will flower well on one side or on some branches but not on the others, or some trees will flower freely while others in the same area and a few yards away have hardly any flowers at all.

It is generally agreed that holly makes the best hedge under English conditions in spite of the many plants introduced from other countries in the last 100 years. Its one drawback is its slow growth. Nevertheless, it is extensively planted for hedges. Unfortunately for the beekeeper, when it is grown as a hedge, periodical clipping prevents flowering. The flowers of the many varieties of holly with variegated leaves or other characteristics that are seen in gardens are probably of similar value for nectar as the wild type and attract bees in the same way.

Some of the American hollies that have been cultivated in this country are good honey plants in their native land, such as *I. opaca* and *I. glauca* (inkberry or gallberry). The latter has been described as one of the best honey plants of the United States, especially in Georgia, the honey being 'light amber, very heavy and very mild and pleasant in flavour' (20).

Hollyhock   *Althaea rosea: Malvaceae*

Flowers of single hollyhocks afford regular feasts for the honey bee as every observant beekeeper knows. It is not unusual to see two or three bees in a single flower at the same time. The pollen is the main attraction and this seems to be of a kind that is particularly to their liking. Late in the season, when pollen is in demand and the number of sources daily diminishing, it is not unusual to find flowers stripped of every vestige of pollen as soon as it appears. When working hollyhock flowers bees often arrive at the hive dusted all over with pollen or bearing a distinct white mark on the back of the thorax as though white-washed. Besides yielding pollen the flowers are also worked for nectar to some extent.

The pollen grain of the hollyhock, often present in the honey of urban or suburban beekeepers, is interesting on account of its large size (100 microns or more in diameter as against three to four microns in forget-me-not).

Various other species of *Althaea* have been observed to be worked for pollen and nectar at Kew, including *A. ficifolia* (Siberia), *A.*

taurinensis (north Italy), *A. cannabina* (Orient) and *A. officinalis* (the marsh mallow).

Honesty   *Lunaria annua* (formerly *L. biennis*): *Cruciferae*

This old-fashioned plant remains popular in gardens both for its sweet-scented purple flowers and silvery flat seed-pods that are used for winter decoration and often coloured. It flowers early, in April or May, and bees are known to visit the flowers for pollen or nectar but not so freely as in most crucifers where the nectar is more easily obtained.

Honey Locust   *Gleditsia triacanthos: Leguminosae*

The fragrant flowers of the honey locust are useful for nectar to beekeepers in its native land (North America) although seldom a source of surplus honey and not so important as the black locust (*Robinia pseudoacacia*—see Acacia). This tree is sometimes to be seen in cultivation and always creates interest on account of the long, branched, woody spines on the trunk. These do not develop as well here as in warmer climates such as the south of Europe, where it is sometimes grown as a hedge.

Honeysuckle   *Lonicera spp.: Caprifoliaceae*

The flowers of the honeysuckle or woodbine (*L. periclymenum*), so prevalent in thickets and hedgerows, and so much loved for their fragrance and beauty, are of little avail to the honey bee because the nectar is out of reach owing to the long, narrow flower-tube. How freely the nectar is secreted every child knows, for who has not revelled in plucking the flowers and sucking the sweet nectar from them. Sometimes the base of the flower is punctured by bumble bees in which case the honey bee may partake of the crumbs of the feast. Most of the climbing honeysuckles seen in gardens also have their nectar out of reach.

Besides the climbing honeysuckles there are a number of shrubby or bush honeysuckles in cultivation. In some of these the flower-tubes are quite short. Among them is the fly honeysuckle (*L. xylosteum*) from Europe which is naturalized in parts of Britain. It has a flower-tube of only some 3 mm. in length, and honey bees are known to frequent the flowers for nectar. Other species that bees visit are *L. standishii*, from China, *L. purpusii* (a hybrid) and *L. tatarica* (from Central Asia). The two first mentioned are particularly valuable for early pollen, for they flower in February. In some seasons at Kew the flowers are humming with bees, whereas in other years the blossoms may be destroyed by frost, or the weather may be against bees visiting them. The Japanese honeysuckle (*L. nitida*), a popular hedge plant, is a useful source of nectar to bees in some countries but in most parts of the British Isles this shrub does not flower.

Hop   *Humulus lupulus: Cannabidiaceae*

The hop plant or vine occurs wild in some parts of England and is grown in gardens as an ornamental climber. It is the parent of the many different varieties of hop that are grown on a large scale for brewing purposes.

The flowers or cones appear in July and are sometimes visited by bees for pollen but are of no special value in this respect. The plant is a wind-pollinated one. In some instances (25) the flowers have been referred to as a source of nectar but there is much doubt in regard to this.

Hop Tree   *Ptelea trifoliata: Rutaceae*

This small tree is wild in southern Canada but has been cultivated in Britain since as far back as 1704. It is sometimes to be seen in gardens, the curious seed-pods always attracting attention. These have been used in home-brewed beer. Small greenish-white flowers appear in clusters in June and July. They are scented and very attractive to honey bees for nectar, the whole tree humming with them at flowering time. The flowering period is short and may barely exceed a week in hot weather. In the eastern U.S.A. the tree is widely distributed and reputed to be a good honey plant in favourable seasons.

Horehound   *Marrubium vulgare: Labiatae*

There are two horehounds, the black and the white, the latter being the better bee plant. White horehound was much esteemed as a medicinal plant in earlier days and always found a place in the herb garden. The dried leaves are still used by herbalists, mainly for making infusions for coughs and colds. In the wild state the plant is not common and usually occurs in waste places. It is a bushy plant with stems 1 to 2 feet high, the stems and leaves being covered with a white woolly down giving the plant a greyish appearance. The small white flowers appear from July to September in dense whorls or clusters on the stem. They are rich in nectar and much loved by the honey bee. The plant is well worth a place in the bee garden, bees often preferring it to other plants.

Black horehound (*Ballota nigra*) is a common plant in hedges and along roadsides. It has wrinkled leaves and purple flowers, also in clusters. Bees visit the flowers but not nearly so freely as with white horehound. The corolla tube is rather long (7 mm.) for the hive bee but the mouth is wide enough for the bee's head to be partly inserted so the nectar may be reached. The plant is common in Europe and naturalized in other countries, including Australia.

Hornbeam   *Carpinus betulus: Carpinaceae*

The yellowish-green catkins are occasionally visited for pollen only, but the tree is of little account to the beekeeper. It may be a minor source of honeydew in some seasons.

Horse-chestnut   *Aesculus hippocastanum: Hippocastanaceae*

The horse-chestnut is perhaps the most commonly planted of the larger ornamental trees in Britain, although a native of south-eastern Europe. It is a good bee plant with its masses of blossoms produced in the latter part of April or May, for these are well worked for pollen and nectar.

Close study of the horse-chestnut in flower will show that the flowers present are of three kinds, male, female and hermaphrodite. Most of the flowers are male with rudimentary ovaries, a few only purely female and the perfect or hermaphrodite flowers chiefly at the base of the inflorescence. The nectar is sometimes clearly visible and is secreted at the base of the flowers between the stamens and the claws of the upper petals. It is protected by woolly hairs. In working the flowers for nectar the honey bee frequently gets at the nectar from the side of the flower or from behind. The nectar guides on the petals are at first yellow but later turn dark red. The flowering period generally lasts for about a month but the petals remain attached and unfaded long after the stamens have withered and the flowers ceased to secrete nectar. It may be observed that bees do not visit such flowers.

Horse-chestnut pollen is very distinctive on account of its bright, almost brick-red colour. Bees returning to their hives with loads of reddish-brown pollen in their pollen baskets when the horse-chestnut is in flower are a common sight, and familiar to most bee-

keepers in or near towns. Sometimes the bees themselves look as
though they have been sprinkled with brick dust as they enter with
their loads.

Various other horse-chestnuts are cultivated and seem to be
equally popular with the honey bee, except of course the double-
flowered forms. The variety *praecox* that comes into leaf and flower
about two weeks before the ordinary form is of interest but is,
unfortunately, probably no longer in cultivation in Britain. The red
horse-chestnut (*A. carnea*, a hybrid) flowers about a fortnight later
than the common horse-chestnut and the blossoms are freely
worked. So also are those of the Indian horse-chestnut (*A. indica*), a
magnificent tree from the Himalayas, not yet much grown in Britain
but quite hardy, at least in the south. It flowers three weeks to a
month later than the ordinary horse-chestnut and so is a source of
nectar and pollen during the early summer or dearth period that is so
obvious to beekeepers in many parts of the country. The Californian
buckeye (*A. californica*) flowers later still (July–August) and the
blossoms are much visited by bumble and hive bees. This species
yields honey in California but is suspected of poisoning or causing
paralysis in bees.

Hyacinth   *Hyacinthus orientalis: Liliaceae*

The ordinary hyacinth, so popular for forcing and for growing
indoors in bowls, is attractive to bees when grown outside in the
flower border. It is in flower early and is visited for nectar and
pollen. The nectar is secreted in an unusual manner. Instead of being
produced at the base of the flower it is secreted in three large drops
from three nectaries appearing as dots near the apex of the ovary.

Hydrangea   *Hydrangea spp.: Hydrangaceae*

The ordinary garden hydrangea (*H. hortensis*) with blue or pink
flowers does not usually attract bees very much, for the bulk of the
flowers in the flower-head are of the sterile type. There are other
hydrangeas, less ornamental perhaps, that are sometimes cultivated
and which are of more use to the honey bee. The best known is
probably *H. petiolaris*, a deciduous climber from Japan, that attaches
itself to walls and trees by means of aerial roots just as ivy does. Its
large, white flower-heads, up to 10 inches across, appear in June
when the inner fertile florets are sometimes freely worked by hive
bees. The plant is useful for covering old tree trunks, walls, mounds, etc.

Hyssop   *Hyssopus officinalis: Labiatae*

This fragrant herb from the Mediterranean region, the oil of which
has been used in perfumery and the leaves in salads, grows well in
England and is sometimes planted as an edging. There are several
varieties of it, with blue, red or white flowers, which are out from
June to October. It looks well with catmint and these may be backed
by lavender and rosemary by the lover of bee plants, thereby providing a useful quartet!

Bees revel in the blossoms, helping themselves to nectar and
pollen. The corolla tube is some 10 mm. in length, but as it widens
into a funnel in the upper part the honey bee is able to reach the
nectar.

Ivy   *Hedera helix: Araliaceae*

The ivy and the honeysuckle are the two most common and widespread British climbing plants. Ivy is to be seen clinging to the
trunks of trees in all classes of woodland except on acid soils and is
very prevalent as ground cover, on rock surfaces and old walls, the
plants sometimes attaining huge dimensions.

The small greenish-yellow flowers of ivy appear very late in the
year, from the latter part of September as a rule until hard weather
appears. They are an excellent source of nectar and pollen to the
honey bee, and where they are prevalent and the weather is warm
enough for bees to work them they may make a welcome contribution
to the hive's winter stores in both honey and pollen. It is the last
important nectar and honey plant of the season to be available to
bees. In mild winters fresh flowers may be found on the plants right
up to Christmas.

As if to make up for the lateness of the flowers and the difficulty
bees might have in driving off the moisture from the nectar and
converting it into honey, the nectar happens to be very concentrated.
It is produced very freely sometimes and may even drip from the
flowers. If insects are excluded, the base of the flower may be covered
with a sugary crust after the flower has faded, so rich is the nectar
in sugar and so lavishly is it produced. The nectar is actually secreted
by a yellowish-green disc surrounding the styles and is freely exposed.
It provides an open feast for all manner of insects besides the honey
bee. Carrion flies are often common visitors, attracted by the strong,
somewhat unpleasant odour of the flowers. Little nectar is secreted

by the freshly-opened flowers but it increases as the flower ages and reaches the female stage.

The pollen of ivy is dull yellow in colour and the individual grain heavily granulated (27). Honey from ivy is said to be greenish in colour with a pleasantly aromatic flavour.

Jack-by-the-hedge   *Alliaria petiolata* (syn. *Sisymbrium alliaria*):
Cruciferae

The white flowers of this well-known wild plant are visited by bees for nectar, which is secreted in four drops at the base of each flower. Known also as hedge garlic on account of the strong smell when bruised this plant is very prevalent in hedges and ditches. Some species of *Sisymbrium* are visited for nectar, such as hedge mustard (*S. officinale*) and flixweed (*S. sophia*).

Jacobaea   *Senecio elegans: Compositae*

These hardy annuals, which are available in so many colours, will attract bees and are generally in flower from May until July if sown at intervals.

Jacob's Ladder   *Polemonium caeruleum: Polemoniaceae*

This bright little perennial is popular in flower borders and rock gardens. It also occurs in the wild state, mainly in the north of the kingdom, but is rare. Bees have been observed industriously working the blue flowers for nectar and pollen. The latter is a bright orange when packed in the bee's pollen baskets. Flowers of other species of *Polemonium* are also visited.

Japonica   *Chaenomeles speciosa* (formerly *C. lagenaria*): *Rosaceae*

The japonica or Japanese quince, also referred to as cydonia japonica, is one of the popular early-flowering shrubs, especially for walls, that are useful to the beekeeper. The blood-red flowers may appear as early as Christmas in some cases but usually not until February or early March. Flowering extends over a long period, often till June. It is mainly for pollen that the flowers are visited but bees have been observed at Kew obtaining nectar in warm weather later in the season. Many varieties of this hardy oriental shrub exist, with white, pink or salmon flowers.

F

Judas Tree   *Cercis siliquastrum: Leguminosae*

This handsome small tree from the Mediterranean region has been cultivated in the south of England for several hundred years, and owes its name to the fact that it is generally supposed to be the tree on which Judas hanged himself after the Betrayal. It grows best in the milder parts of the country, flowering between April and June as, or before, the young leaves appear. It is then very picturesque, the purple flowers being borne in profusion, often in clusters on the old wood. These may be used in salads.

Bees visit the flowers freely for nectar but the trees are never sufficiently common to be of much consequence. An allied species (red-bud; *C. canadensis*) is common in the south-eastern United States and is a good early bee plant there, but does not provide surplus honey (19).

Juneberry   *Amelanchier leavis* (formerly *A. canadensis*): *Rosaceae*

The Canadian juneberry or serviceberry has been cultivated in Britain for 200 years and is naturalized in some areas. It usually flowers in April and is then a mass of white, the flowers being followed by edible fruits ripening in June. While the blossoms may be visited for pollen they do not appear to be especially attractive to bees or to be worked for nectar as a rule. In its native land it seems to be of some nectar value.

Kalmia   *Kalmia spp.: Ericaceae*

These shrubs, which are mostly evergreen and like peaty, moist soils, are often to be seen in gardens and have become more or less naturalized in some instances. The species most generally grown is the sheep laurel (*K. angustifolia*) which may form thickets 15 feet across through sucker growth and bears clusters of rosy-red flowers in June. This and the calico bush (*K. latifolia*), also much cultivated and one of the most beautiful of evergreen shrubs, are both considered to be useful nectar plants in their native land (eastern North America), although the latter has been credited with being a source of poisonous honey.

Koelreuteria   *Koelreuteria paniculata: Sapindaceae*

This large tree from northern China is not often seen in cultivation. It produces panicles or loose bunches of yellow flowers at the ends of the branches in July and August and these may be heavily worked for nectar by honey bees.

Laurel   *Prunus laurocerasus: Rosaceae*

The cherry laurel, one of our most useful and quick-growing ever-greens, is much visited by bees at times. It flowers in April but bees may be seen visiting it more or less at any time, particularly when ordinary nectar is scarce. This they do for the sake of the sweet fluid secreted from the extrafloral nectaries on the undersurfaces of the leaves, particularly the young growth. Bushes may sometimes be found humming with bees working in this way. The writer has found this very noticeable in early June in an area near Bagshot Heath where there has been little other bee fodder available at this time.

The Portugal laurel (*P. lusitanica*), which flowers later, and the bay laurel (*Laurus nobilis*) are also visited by bees. The latter is well known for its sweet-scented leaves often used for flavouring milk puddings. Its small greenish-yellow flowers appearing in May or June are rich in nectar and attract the hive bee (4).

Laurustinus   *Viburnum tinus: Caprifoliaceae*

This much-grown ornamental shrub flowers mainly in the winter months, between autumn and early spring. Its pinkish-white flowers are fragrant and in dense bunches. On fine days in early spring they may be visited for pollen by bees but do not seem to attract much attention later in the year. In view of its prevalence in southern gardens this well-known evergreen is doubtless a useful early pollen source to urban beekeepers. The pollen is of a pale slate-grey colour in the bees' pollen baskets.

Lavatera   *Lavatera spp.: Malvaceae*

There are annuals, biennials, and perennials among the lavateras or tree mallows. All have the same type of flower structure and yield pollen in abundance which honey bees readily make use of. They also obtain nectar. In the milder districts *L. arborea* is one of the best of the taller perennial kinds and an imposing plant. It grows to 10 feet in height.

Lavender   *Lavandula spp.: Labiatae*

It is common knowledge that the fragrant blue flowers of lavender are always a great source of pleasure to the honey bee and that much joyful buzzing always accompanies their visits. June and July are the months when the flowers are at their best in most districts. The flowers produce nectar freely, which is stored at the base of the flower and protected by a ring of hairs. The flower-tube of the usual garden lavenders is about 6 mm. long—a suitable length for the hive bee.

Lavender needs to be grown in full sun and does best on light or chalky well-drained soils. These are the conditions in many parts of southern Europe where the plant occurs wild over large tracts of country and supplies good-quality honey. This has been described as dark, of very pleasant flavour, and granulating with a grain almost as smooth as butter (*American Bee Journal*, 1937, 412). Along with the orange and rosemary it is one of the three most important honey plants in many districts in Spain.

Lavender is sometimes grown on a field scale in England (e.g. Hitchin, Herts) for perfumery purposes, and should afford opportunities for nearby beekeepers. Unfortunately, the flowers have to be harvested while still in their prime.

Lilac   *Syringa spp.: Oleaceae*

The sweet-scented flowers of this favourite ornamental shrub secrete nectar freely but in most varieties the flower-tube is too long for the honey bee. When nectar is produced abundantly it may rise from 2 to 4 mm., or even more in the flower-tube, and so become within reach of the hive bee. Under such conditions honey bees may be seen working lilac. Honey bees have been observed working the masses of pink flowers of *S. tomentella* (from western China) at Kew on occasions. There are also reports that *S. reflexa* (from central China) has been much worked in parts of Germany (*Bee World*, 1938, 62).

Lily   *Lilium spp.: Liliaceae*

Lilies are essentially lepidopteroid flowers, being often visited by moths at night. Pollen is produced abundantly and is often bright coloured. Occasionally honey bees make use of it.

Lily-of-the-Valley   *Convallaria majalis: Liliaceae*

The honey bee is known to visit these sweet-scented flowers but only for pollen (11). They appear in May and June when there are many other, possibly more appetizing, sources of pollen available. The plant occurs wild in most parts of Europe as well as in Britain.

Limnanthes   *Limnanthes douglasii: Limnanthaceae*

This Californian annual has long been grown in English gardens and
its attractiveness to bees when grown in the mass realized. It is also
invariably included in the collections of seeds of bee plants that are
sometimes listed in seedsmen's catalogues. If a few odd plants only
are grown in the flower border they often prove disappointing and
bees do not visit them. The plant has a somewhat straggling habit
but is quite well suited for edgings, beds and rockeries. Any soil suits
it but a moist situation is preferable. For early spring-flowering seed
should be sown in September and for summer-flowering in March.

The yellow and white flowers are sweet scented and varieties
differing in size and colour exist. The names butter and eggs and
meadow foam have been applied to them. Under favourable conditions they produce nectar freely and are well liked by bees. It is
considered a good nectar plant, especially for stimulative purposes,
in its native land (23).

Loganberry   *Rubus hybrid.: Rosaceae*

The loganberry was introduced to Great Britain in 1900 from the
United States where it had been raised by Judge Logan of Santa
Cruz, California, in 1881 by crossing the wild dewberry with a
raspberry. Through seedling plants it has not always kept true to
type and other berries like the phenomenal berry and laxtonberry
have been derived from it. With proper management the loganberry
will succeed on all manner of soils but gives the highest yields on
deep, rich loams.

The flowers are rich in nectar and resemble those of raspberries
with which they are probably on a par as bee fodder, for they are
frequented continually by bees in suitable weather during the period
of several weeks when they are available in the early summer.

Loosestrife   *Lythrum salicaria: Lythraceae*

The purple loosestrife is one of the most handsome of British wild
plants and is to be seen in all parts of the country, especially where
soil conditions are moist as on the edges of pools, streams and
ditches. In such situations it often forms a vivid patch of colour on the
landscape. It is 2 to 4 feet in height with flowering spikes about 1 foot
long bearing rich purple-red flowers from June to August. The flowers
are of interest in that the length of stamens and styles varies in

different plants as does the pollen from them. This may be yellow or greenish with large or small individual grains.

Purple loosestrife is a good bee plant and supplies nectar and pollen in quantity. Unfortunately it is seldom sufficiently abundant for surplus in Britain but the honey is said to be dark with a strong flavour (27). The plant is plentiful in most parts of Europe and is widely naturalized elsewhere, e.g. Australia and North America. There are also many garden forms of it with flower spikes larger and of a different shade. These are equally attractive to bees.

The yellow loosestrife (*Lysimachia vulgaris*), an entirely different plant, is only of interest to the honey bee for pollen.

Lucerne   *Medicago sativa: Leguminosae*

Lucerne, or alfalfa as it is sometimes called, is not the important crop in Britain that it is in other countries. It has, however, been grown to a greater extent in recent years both for hay and green fodder. This has been mainly in the southern counties, especially in chalk districts, as the plant favours lime soils.

This clover-like plant is perennial and once established yields several cuts in a season, the plants lasting for many years. In the moist conditions of Britain the plant appears to make more leafage than it does in other countries and is cut before flowering, whether for hay or fodder, otherwise the plants become too fibrous. Sometimes three or even four cuts may be made throughout the summer, all before any flowers at all have a chance to appear. So the crop may be quite useless to the beekeeper. In some cases, however, there are flowers available to bees later in the season and these may prove a useful source of nectar, particularly towards the end of hot dry summers (14).

Self-sown lucerne plants are sometimes to be seen in flower by the sides of paths and fields. Their purple flowers do not appear to attract bees much until later in the year when other sources of nectar become scarce.

Lucerne or alfalfa honey is known to vary much according to locality but is of good quality and generally resembles clover honey but with a more spicy flavour and tendency to granulate early.

Other plants allied to lucerne that the honey bee visits for nectar are Cossack alfalfa (*M. media*), Siberian or yellow lucerne (*M. falcata*), toothed medick (*M. polymorpha*, formerly *M. denticulata*)—rather rare—and black medick or yellow trefoil (*M. lupulina*). The last mentioned is widely distributed in Britain, being most prevalent on limestone soils. Seed is often included in seed mixtures for grassland, especially for temporary leys on light or chalky land. Flowering is generally earlier than is the case with white or red clover, the small yellow flower-heads being followed by distinctive black seed-pods. The flowers are often very freely worked for nectar.

Lupin   *Lupinus polyphyllus: Leguminosae*

The flowers of most cultivated lupins are generally regarded as nectarless (11), honey bees and bumble bees visiting them mainly for pollen, the latter with its greater weight being better able to manipulate the flowers. The flowers are of little account as far as the beekeeper is concerned.

Magnolia   *Magnolia spp.: Magnoliaceae*

Magnolia flowers are not conspicuously attractive to bees. They are of course only to be seen in gardens and are probably visited mainly for pollen. It is interesting to observe that two of the North American species commonly grown in Britain (*M. grandiflora* and *M. virginiana*, formerly *M. glabra*) are reputed to be useful sources of nectar in their native land (19).

Mahonia   *Mahonia aquifolium: Berberidaceae*

This low-growing evergreen shrub, with its holly-like leaves, is much cultivated on account of its ability to withstand shade and is naturalized in some parts of the country. Its grape-like fruits, which may be used for jelly, although bitter, account for its other common name—Oregon grape. It is in flower for several weeks commencing early in April and its small yellow blossoms are worked moderately by bees for nectar.

Maize   *Zea mays: Gramineae*

Maize or Indian corn is rarely grown in England except for silage although sweet corn, which is but a form of it, is more extensively cultivated than formerly. The male flower-heads or tassels furnish an abundance of pollen which is sometimes collected by bees. Usually, however, there are numerous other (probably more palatable) sources of pollen available when they appear and they are ignored. There have been cases of bees collecting insect honeydew from the plants (23). There also seems evidence that under some conditions (rapid growth) the leaf sheaths may split exposing a certain amount of sweet sap or juice which bees collect (*American Bee Journal*, 1936, 82).

Mallow   *Malva sylvestris: Malvaceae*

Although just a weed the common mallow is a handsome plant with its large mauve flowers. These are to be seen at their best from June to August when they always find favour with bees, not only for pollen but for nectar also. The pollen is white or very pale mauve in colour. Bees often get themselves covered in it and the top of the thorax quite white as if daubed with white paint. The pollen grain under the microscope is spherical, rough and exceptionally large, as is not unusual in members of this family (see Hollyhock). Its diameter may be as much as 144 microns (11) as against 3 to 4 in forget-me-not, 8 in goat's rue (galega) and 30 to 35 in white clover.

The musk mallow (*M. moschata*) and dwarf mallow (*M. neglecta*, formerly *M. rotundifolia*) are other wild species that are useful to bees and are in flower until frosts arrive. The former is a particularly handsome plant with rosy-pink flowers and is sometimes listed among collections of bee plants in seedsmen's catalogues. A white-flowered variety of it is also cultivated.

Malope   *Malope trifida: Malvaceae*

This showy annual that originated in southern Spain attracts bees. It is best sown where it is intended to flower and well thinned. The varieties generally cultivated ('Grandiflora') have white, crimson and flesh-coloured flowers.

Maple   *Acer spp.: Aceraceae*

There are about a hundred different species of maple distributed more or less throughout the northern hemisphere. As a group they are good bee plants, producing nectar freely and supplying early pollen. The open nature of the flower enables the honey bee to have easy access to the nectar, but as most maples flower very early in the season (March–April) the extent to which bees are able to make use of them is largely dependent upon suitable bee weather at the time.

The only maple native to Britain is the field maple (*A. campestre*) often to be seen as a small tree on calcareous soils and common in hedges, where it may be coppiced. When allowed free growth it often reaches 50 to 60 feet in height. Its bunches of upright, delicate green blossoms appear in April or May and produce nectar freely. This is secreted by a thick, fleshy central disc and is freely exposed as in other maples. If the tree were more common it would doubtless be valuable to the beekeeper, as is the sycamore (see Sycamore).

A number of different maples are frequently to be seen in cultivation as ornamental trees, usually for their foliage, which may be attractively cut or give yellow or reddish tints in autumn. Among the maples that have been observed by the writer to be worked for nectar the following are among the better known: the Italian maple (*A. opalus*), very free flowering and often grown as a street tree; Norway maple (*A. platanoides*); Montpelier maple (*A. monspessulanum*); Oregon maple (*A. macrophyllum*); box elder (*A. negundo*); and sugar maple (*A. saccharum*). The last three are natives of North America where they are valuable for early nectar and pollen in many areas (19).

Marjoram *Origanum vulgare: Labiatae*

Wild marjoram is one of the best known and best loved of English wild flowers with its delightful fragrance and masses of purple flowers. It favours calcareous soils and is very common on chalk hills in the south of England, often dominating the landscape when in flower. This takes place from July onwards. The flowers are in crowded clusters at the ends of the stalks, which vary from 1 to 2 feet in height. Honey bees are frequent visitors for nectar and this is yielded abundantly at times. Honey is not usually obtained pure from this plant in England as there are generally other nectar-yielding flowers available at the same time, such as the blackberry and wild thyme. However, it is considered to yield a high-quality honey of good flavour and aroma and its presence is always likely to improve rather than detract from the quality of other honey, particularly that of willow-herb which is naturally a honey of little flavour (14). Willow-herb is in flower at the same time as marjoram.

In gardens it is sweet or knotted marjoram (*O. majorana*) and pot or perennial marjoram (*O. onites*) that are generally cultivated, their aromatic leaves being used either in the fresh or dried state for flavouring. They are much used for soups and for stuffings. Sweet marjoram is a native of North Africa, and is grown as an annual in Britain as it does not withstand the winter. Plants grown from seed sown in April usually flower in July. The flowers are small and pale, with the flower-tubes about 4 mm. long. They are much sought after by bees for nectar.

Marsh Marigold   *Caltha palustris: Ranunculaceae*

The marsh marigold is common in moist pastures and by the sides of streams. It always prefers damp situations as the name implies. The flowers, which might be those of a giant buttercup, are among the first of the wild flowers and sometimes appear as early as February, continuing until June. Honey bees are frequent visitors for pollen and nectar, the latter being half concealed and secreted, sometimes quite abundantly, in two shallow depressions at the base of the flower (11). There are garden forms of the plant, some with double flowers.

Mayweed   *Anthemis cotula: Compositae*

The mayweed, or stinking chamomile as it is sometimes called on account of its strong odour, is not considered of much account as a bee plant in Britain, in spite of its prevalence, for it is one of the commonest weeds. However, it is interesting to note that in parts of the United States, where it has become an equally common weed, it is esteemed by beekeepers and even credited with giving surplus honey, which is described as light yellow and very bitter (20). Its value there is largely on account of its flowering during dearth periods— between spring and summer.

Meadow Saffron   *Colchicum autumnale: Liliaceae*

An uncommon wild plant, meadow saffron or autumn 'crocus' is cultivated for its medicinal value. It causes poisoning in cattle. Bees are known to visit the large purple flowers that appear in late August or September for pollen.

Meadowsweet   *Filipendula ulmaria* (formerly *Spiraea ulmaria*):
Rosaceae

The dainty, sweet-scented flowers of this very common meadow plant are often visited by bees for pollen. They are common from June to August. When working the flowers for pollen the honey bee returns with her pollen loads distinctively greenish in colour. There is doubt whether the flowers are also a source of nectar in Britain, but in Europe bees have been observed working them for nectar (11) (*Bee World*, 1939, 83).

Medlar   *Mespilus germanica: Rosaceae*

Although a much-prized fruit with our forefathers, room is seldom
found for the tree in present-day orchards and gardens. Some fine
old trees, picturesque with their crooked branches, may often be seen
in old gardens. The large white flowers that appear in May are not
unlike those of the apple and attract the honey bee for nectar and
pollen. The nectar is secreted by a fleshy yellow ring at the base of the
flower. A wild form of the medlar exists in thickets but is rare. The
cultivated tree is nowhere sufficiently common to be of any consequence to the beekeeper in Britain but is of more account in other
countries, e.g. Portugal.

Melilotus   *Melilotus spp.: Leguminosae*

Melilotus or sweet clover, also known as Bokhara, bee, or honey
clover, is well known to most beekeepers. The very name of course
suggests honey (*mel*—honey; *lotus*—flower). This is appropriate for
it is always attractive to bees and an important honey-producing
plant in many parts of the world, and has been renowned as a bee
plant from classical times.

There are two kinds of sweet clover commonly seen, the white
(*M. alba*) and the yellow flowered (*M. officinalis*). They are considered to be of equal value as bee plants. Both occur wild or as
occasional weeds in Britain. They may also be grown as decorative
plants in gardens or for fodder or hay. Unfortunately for beekeepers,
however, sweet clover has never been extensively grown as a farm
crop in the United Kingdom, objections to it being its somewhat
fibrous or woody nature and the fact that stock do not take to it
readily owing to its bitterness and require to be educated to it.
However, these objections must apply in other countries, but have
not prevented its becoming a major agricultural crop in them.
Sweet clover has, however, proved useful as a green manure on thin,
sandy soils (e.g. the Breckland), the decay of its large fleshy roots
contributing to soil fertility apart from the stems and leafage.

Both white and yellow sweet clover are biennials and do not flower
until the second year. They then have a long flowering season. This
commences in June or July, when numerous flowers appear on the
tall, much-branched stems. These are rich in nectar and invariably
swarm with bees. The white-flowered species (*M. alba*) is the one
usually grown for agricultural purposes. It is the taller and more

vigorous plant, producing more leafage. It flowers about a fortnight later than the yellow but does not stand cutting back so well. Both grow readily in all types of soil, except the very acid, and often thrive where few other plants will grow, such as the banks of quarries, railway embankments, etc. White sweet clover usually grows 5 to 6 feet high but may reach 8 to 10 feet under good conditions.

The honey from sweet clover is of good quality, being light in colour although often greenish. It is of medium density, with a pleasant mild flavour, slightly vanilla-like. Granulation takes place more readily than with ordinary white clover (*Trifolium repens*). The honey is known to vary from different districts and has been accused of causing mild headache in some individuals owing to the presence of coumarin. The pollen is very like that of ordinary clover.

The history of white sweet clover in the United States is of interest. It first appeared as a weed and was actually scheduled as a noxious weed in some States. Later it began to be grown to a small extent as a pasture and hay crop, beekeepers being known to scatter seeds on the edges of fields and along roadsides, surreptitiously and by night so we are told (19)! Eventually it was cultivated on an intensive scale in the mid-West region and became the source of vast quantities of honey, regions of hot dry summers being best suited for honey production. It was even the cause of derelict farms and land being restored to prosperity. In more recent years, however, its cultivation has declined owing to pests and diseases, particularly the sweet clover weevil. Other crops have now taken its place over wide areas to the detriment of beekeeping.

Many varieties of sweet clover have appeared from time to time and new ones are likely to arise as a result of breeding work being done on the crop. One of the first to arouse interest, about a quarter of a century ago, was Hubam clover, an annual form of white sweet clover. Agricultural trials with it in England were disappointing for it produced no more leafage than other clovers, was less robust than the ordinary sweet clover and subject to attack by the turnip flea beetle in the seedling stage. It was also inclined to revert to the original biennial form. Other varieties that have attracted attention are 'Alpha' sweet clover, which grows about 4 feet high with numerous thin stems (about thirty) instead of a solitary thick stem, and 'Melana' sweet clover, an annual form of it (*American Bee Journal*, Dec., 1932). The last mentioned has grown well with the writer, remaining in flower for a long period and proving very attractive to bees.

Another *Melilotus* sometimes found in waste places in Britain is *M. indica*. It is cultivated in some countries as an agricultural crop or orchard cover and is a good nectar plant. It is an annual and often known as King Island melilot. In parts of Australia it has become a weed and caused trouble through the seeds tainting or giving a hay-like smell to bread when they occur as an impurity in wheat. Milk and cream have also been affected by the plant.

Mesembryanthemum   *Mesembryanthemum edule: Aizoaceae*

This Cape plant is now naturalized on cliffs along the coast in some parts of Devon and Cornwall and is often to be seen in cultivation at South Coast seaside towns to cover and maintain banks and other exposed places. It soon carpets the ground with its creeping stems and fleshy leaves. The large showy flowers only open in the sunshine when they attract honey bees in numbers for nectar and pollen. The closely allied ice-plant (*M. crystallinum*), often cultivated, has been recorded as a good bee plant in warmer climates where it grows freely out of doors. Honey from it is said to granulate very quickly (23).

Micromeria   *Micromeria rupestris: Labiatae*

The small white flowers of this low-growing plant from southern Europe are a great attraction to bees for nectar from July onwards. Having no special horticultural charm it is not in general cultivation. With its prostrate stems, which turn up at the ends when in flower giving a heath-like effect, it is suited for the rock garden. The leaves are scented and with its abundance of late flowers it is worth a place in the bee garden. It is a good bee plant at Kew and the same has been recorded of it elsewhere (*American Bee Journal*, 1933, 393).

Mignonette   *Reseda odorata: Resedaceae*

Few garden plants can compare with mignonette for fragrance and few can excel it for attracting bees when in full bloom on a fine morning in a good-sized bed. For continuous blooming seed should be sown at intervals in the spring and summer. Besides the common kind a selection may be made from the coloured (red and yellow) and dwarf forms. Nectar is the main attraction but pollen is collected also. Bees visit the flowers all day long. Some think mignonette is capable of giving more blossom and more nectar for a given area than any other plant.

The wild mignonette (*R. lutea*) which is such a common plant on chalk hills and in flower from June to August is also a good source of nectar and pollen to the honey bee, but is probably nowhere sufficiently abundant for surplus. Unlike its garden counterpart, that originated in Egypt, the flowers have little fragrance.

Milkweed   *Asclepias syriaca: Asclepiadaceae*

Although not generally cultivated in Britain this economic plant is of interest to beekeepers. Besides being a good source of nectar it is also a potential yielder of seed-floss (kapok substitute), fibre, and even rubber. It has been grown for seed-floss in the United States. The plant grows readily in Britain, reaching 4 to 5 feet in height and flowering freely. Like other members of the family the pollen is in two masses (pollinia) connected by a strand. These strands may become entangled in bees' legs and in allied species in central Europe are believed to be the cause of heavy mortality among honey bees (*Bee World*, 1937, 71). Good honey has been obtained from this source.

Milkwort   *Polygala vulgaris: Polygalaceae*

The common milkwort, often so prevalent on chalk downs with its purplish flowers in May and June, is reputed to be visited by honey bees but is probably of little consequence as a bee plant. Its pollen grain is unusual in shape and beautifully marked (27).

Mint   *Mentha spp.: Labiatae*

The various mints, wild and cultivated, are all good bee plants and yield nectar freely. It is rarely however that they occur in sufficient abundance to yield surplus honey. The fields of mint that are grown for culinary purposes are normally harvested before they flower, which is unfortunate for the beekeeper. At least eight different kinds of mint are grown commercially in Britain, many of them hybrid forms. Among the wild mints two of the most common are water mint (*M. aquatica*) and field mint (*M. arvensis*), also known as corn mint. Both flower late in the season, from August onwards, and are a useful source of late nectar for bees in districts where they grow freely. The former sometimes occurs in masses near streams, as in parts of East Anglia, and the latter is common in corn stubble after harvesting. Pennyroyal (*M. pulegium*) with its more straggling habit may occur freely in and around bogs. It is often to be seen in gardens, especially rock gardens, in one or other of its many forms. Penny-royal has become naturalized in other countries and is a source of honey in New Zealand where it has spread to the extent of becoming a weed. The honey from it is pale in colour and rather thin. Spearmint (*M. spicata*) and peppermint (*M. piperita*) are sometimes cultivated on a field scale and surplus honey has been obtained from them. It is amber in colour with a minty flavour when fresh but this becomes less noticeable in time (23).

Mock Orange   *Philadelphus spp.: Philadelphaceae*

The highly-scented flowers of the mock orange are known to attract honey bees for nectar and pollen (11) but not as a rule in large numbers. There are many different kinds in cultivation, some with double flowers.

Motherwort   *Leonurus cardiaca: Labiatae*

This plant is sometimes to be seen apparently wild, but is neither common nor indigenous in Britain. It is prevalent in parts of Europe and is regarded there as a good honey plant. One writer refers to its long flowering season, from July onwards, and considers it better than phacelia, borage and sweet clover as a nectar plant (*Leipziger Bienenzeitung*, Feb., 1933). It reaches 3 feet in height, has character-istic palmate lower leaves and white or purplish-pink flowers on the stem. The flower-tube is only some 4 mm. long and the pollen white.

Motherwort is now extensively naturalized as a weed in other countries, especially Canada and the U.S.A. where it is also very attractive to bees (19). Nectar may be secreted very freely.

Mountain Ash   *Sorbus aucuparia: Rosaceae*

The rowan or mountain ash is a characteristic wild tree of the north and west of Britain. It is also much planted as an ornamental tree on account of its feathery foliage. The bunches of greenish-white flowers generally appear in May and June and are visited by the hive bee for nectar and pollen, but not with much enthusiasm or zest as a rule.

Myrobalan   *Prunus cerasifera: Rosaceae*

The myrobalan or cherry plum is a well-known small tree in gardens and is sometimes used as a hedge plant or as a stock for grafting plums. It is among the first deciduous trees to flower and is prized on this account, trees being covered with the pure white blossoms in March. The purple myrobalan (cultivar 'Pissardii') which originated in Persia is similar but has purple leaves and pale rose-coloured flowers. It is very commonly cultivated on account of its foliage—often as a street tree. The flowers of both these cherry plums are worked by the honey bee for nectar and pollen when weather is suitable and are a useful early source of provender. They may be in flower even before the almond—in February in early seasons.

Myrtle   *Myrtus communis: Myrtaceae*

Although so well known for its fragrance and use in bridal bouquets this 'classical' plant is only hardy in the milder parts of the country. It is often to be seen as an evergreen shrub in gardens in Devon and Cornwall and produces white flowers in late summer. These yield pollen and perhaps a little nectar for the honey bee.

Nasturtium   *Tropaeolum spp.: Tropaeolaceae*

The many dwarf and climbing forms of this popular garden annual may be visited by hive bees, but only for pollen, the nectar being out of reach.

Nemophila   *Nemophila spp.: Hydrophyllaceae*

Among the most easily grown of annuals, the nemophilas thrive in any soil. They are well suited for edging and small beds, being of compact growth. The best displays are probably obtained in the north and the cooler parts of the country. The flowers, with their many shades of blue, are favourites with the honey bee for nectar. Autumn and successional spring sowings give a longer flowering period.

Nettle   *Urtica dioica: Urticaceae*

The common stinging nettle, with its small inconspicuous flowers, is of no interest or value to the bee, but may interest the beekeeper for nettles have been advocated for obstructing the entrances of hives that are being robbed, the effect being said to be almost like a continual wetting until the leaves wilt (*Bee World*, 1932, 132).

New Zealand Flax   *Phormium tenax: Liliaceae*

Attempts have been made to grow this plant on a commercial scale for fibre in England but without much success. However, it is often to be seen growing as an ornamental plant. Bees visit the flowers for nectar.

Nigella   *Nigella spp.: Ranunculaceae*

The nigellas (love-in-a-mist) are yet another group of popular garden annuals that are visited by bees for nectar and pollen, but usually only in a moderate degree. The flowers are interesting in having an unusual type of nectary and nectar secretion.

Niger Seed   *Guizotia abyssinica: Compositae*

The plant yielding niger seed, a small black seed used in bird seed mixtures and in some countries as an oil seed, appears to be a good bee plant. Its flowers are well worked for nectar at Kew in the late summer. The plant is not in general cultivation.

Nipplewort   *Lapsana communis: Compositae*

This is a common plant in some areas, by hedges and roadsides, and produces its pale yellow flower-heads, rather like those of the sow-thistle, from July onwards. Bees sometimes visit them, particularly late in the season when other bee forage becomes scarce.

Oak   *Quercus spp.: Fagaceae*

The wild and cultivated oaks are of little consequence as bee plants, although bees do sometimes collect pollen from them. This is produced in April or May, when many other sources are available. There are reports of nectar from the oak, but this has probably been insect honeydew. The oak is the host plant of a very large number of different kinds of insect, including many gall insects. In some years it becomes a troublesome source of honeydew, especially late in the season.

Olearia   *Olearia x haastii: Compositae*

This bushy evergreen shrub is often seen in gardens and is one of the few New Zealand shrubs that are hardy in Britain. It flowers in July and August. The flower heads are visited by bees for nectar and pollen, but only to a limited extent. The shrub is to be seen at its best in coastal districts.

Oleaster   *Elaeagnus spp.: Elaeagnaceae*

The sweet-scented blossoms of several different oleasters have been observed to be much frequented by hive and bumble bees in June. However, it is doubtful whether the former are able to obtain much nectar from the flowers of these shrubs owing to the length of the flower-tube.

Olive   *Olea europaea: Oleaceae*

The olive is little more than a curiosity in Britain and can only be grown in the mildest parts or with the protection of a wall in the south. In Spain it is a well-known source of honey.

Orchid   *Orchis spp.: Orchidaceae*

There are records of the honey bee visiting wild British orchids (e.g. *Orchis latifolia* and *O. morio*). Some orchids secrete nectar in abundance, but often it is out of reach of the hive bee. In any case they are not sufficiently plentiful to be of any consequence to the beekeeper.

Osmaronia (formerly Nuttalia)   *Osmaronia cerasiformis: Rosaceae*

Although not generally cultivated, this deciduous shrub (formerly *Nuttalia cerasiformis*) is of interest on account of its early flowering (February–March) and the zeal with which bees work its white almond-scented blossoms for pollen when opportunity occurs. These are produced in great profusion. The shrub eventually forms a thicket several feet across and 6 to 8 feet high with numerous stems from the base. It is quite hardy in Britain although a native of California, where it is called oso berry. Male and female flowers are on different plants, the latter producing plum-like fruits which are purple when ripe. The flowers are a source of nectar as well as pollen.

Oxydendrum   *Oxydendrum arboreum: Ericaceae*

For those situated in heath or peaty areas this small tree from the eastern United States may be of interest for it thrives under the same conditions as azaleas and rhododendrons. Bunches of white flowers are produced in July and August. These are an important source of high-quality honey in its native home, where it is known as sourwood or sorrel tree, owing to the acidity of the leaves (20).

Pagoda Tree   *Sophora japonica: Leguminosae*

The so-called Japanese acacia or pagoda tree (*S. japonica*) is a handsome tree of Oriental origin sometimes cultivated for ornament in Britain and elsewhere. It bears masses of sweet-scented cream-coloured flowers in September, which fall while still fresh, literally carpeting the ground. Flowering lasts for about a month and the flowers are actively worked by honey bees for both nectar and pollen. They must constitute a useful late food source for the bees. In good flowering seasons at Kew trees have been observed alive with bees, the busy hum being audible several yards away, bees even buzzing about the carpet of fallen flowers.

The tree is one of the last to blossom in the year. Unfortunately it is rather uncertain as a honey plant in Britain, for after poor or wet summers it hardly flowers at all. Another drawback is that it does not commence to flower until thirty or forty years of age. It has been recorded in Europe that the flowers poison bees, and that beekeepers at Marienfeld in the Banat are in the habit of moving their hives from the vicinity of the trees at blossoming time to avoid mortality (*Bee World*, 1923, 7). No sign of poisoning has been observed at Kew during several seasons' observation, which suggests that this may be climatic. Honey from the tree in France is said to have a pronounced flavour (4).

Other species of *Sophora* are worked by bees, notably *S. viciifolia* —a handsome June-flowering, prickly shrub, 6 to 8 feet in height. It is also a native of China where it covers extensive tracts of barren country just as gorse does in Britain.

Parsnip   *Pastinaca sativa* (formerly *Peucedanum sativum*):
Umbelliferae

Both the wild and the cultivated parsnip of the vegetable garden are known to be sources of nectar to the honey bee. The wild plant has a tough fibrous root quite unlike the large fleshy root of its cultivated relative, but the leaves and flowers are similar. It is abundant in many parts of the country, and is to be seen around fields, in hedgerows and meadows, and on sea cliffs.

Its bunches of yellow flowers are present from July to September, and are visited by all kinds of insects, including the honey bee. It would appear that nectar is not usually present in abundance, and that it is unusual for it to be of any consequence as a nectar source. One beekeeper in Surrey relates how a 20-acre fallow field at the back of his house, which was covered with the plant, gave a good crop of honey in one year, but in the four previous years he had hardly seen any bees on it (*Bee World*, 1931, 114).

Where the garden parsnip has been grown on an extensive scale for seed purposes, crops of honey have been obtained. This has been described as light amber in colour, but not of the best quality, with a flavour slightly suggestive of the source from which it was obtained and as being inferior to that from celery. The pollen is greenish-yellow and the individual grain long and narrow (27).

Pea   *Pisum sativum: Leguminosae*

The flowers of the garden pea are not of much interest to the honey bee as a rule, but in some instances they are believed to be a source of nectar. Possibly the variety grown and the size of its flower may have some bearing on the matter.

Pea-tree   *Caragana arborescens: Leguminosae*

The Siberian pea-tree is a large deciduous shrub 10 to 15 feet in height. It is often cultivated and there are many garden varieties. In May or June it bears yellow pea-like flowers singly on the stems. At Kew the flowers are attractive to bumble bees, but it is not often that hive bees are seen in their vicinity. It is reputed to be a good bee plant in Canada (Ontario), and to be well suited for poor soils and as a hedge plant (*Canadian Bee Journal*, Dec., 1940).

Peach   *Prunus persica: Rosaceae*

Like its near relative the almond, this fruit tree flowers early and is a source of nectar and pollen. It is of some importance to beekeepers in warmer climates, but in Britain is only to be seen occasionally on walls or in glasshouses. The flowering peaches seen in gardens usually have double flowers which open two to three weeks later than the almond. Bees have been successfully used as pollinators in peach houses in some parts of the country by large growers, thereby effecting a big saving in labour as compared with hand-pollinating or rabbit-tailing and resulting in more effective pollination (*Bee Craft*, 1940, 7).

Pennycress   *Thlaspi arvense: Cruciferae*

The small white flowers of this annual weed are visited by bees for nectar (11). It is often common in cultivated ground and is to be found in all parts of the country.

Penstemon   *Penstemon spp.: Scrophulariaceae*

Bees visit penstemons to some extent. They are mostly of hybrid origin and with flowers of various colours and sizes. *P. antirrhinoides*, which is generally grown against a warm wall, being not altogether hardy, is reputed to be a source of surplus honey in its native land—California. Several of the wild penstemons of North America are known to be good bee plants (*American Bee Journal*, 1925, 526).

Peony   *Paeonia spp.: Ranunculaceae*

Most of the ponies seen in gardens have 'double' flowers and are of little interest or use to the honey bee. The single-flowered sorts, however, produce an abundance of pollen and bees may often be seen collecting it, sometimes three or four on a flower together. It has been estimated that as many as three million pollen grains may be produced by a single peony flower (20).

Perezia   *Perezia multiflora: Compositae*

This plant, from the slopes of the Andes, is little known in cultivation but is very attractive to bees. It is an annual or biennial, reaching 3 to 4 feet in height with masses of china-blue flowers with yellow centres rather like a Michaelmas daisy. As many as 100 flower-heads may be carried by a single stalk. The flowering period lasts for several weeks, usually from early June onwards. Seeds are produced in abundance and seedlings come up freely round the old plants by the time autumn arrives. These generally survive the winter and only need thinning in the spring. Bees work the flowers for nectar at all hours in fine weather and the masses of pollen in their pollen baskets are conspicuous on account of their pure white colour. The plant is a useful and easily grown one for the bee garden.

Perovskia   *Perovskia atriplicifolia: Labiatae*

It is surprising that this exceedingly handsome Himalayan shrub is not more generally grown in English gardens, for it is quite hardy, although often dying back in winter and sending up fresh flowering shoots the following season. Flowering takes place in August or September when the masses of purple lavender-like blossoms are covered with bees probing for nectar. The blossoms contrast well with the silver grey foliage of the plant. It generally reaches about 6 feet in height, needs full sun and thrives best on light well-drained soils. The extent of flowering is dependent a good deal on the season, fine hot summers giving the best results.

Phacelia
Phacelia tanacetifolia: Hydrophyllaceae

This annual is well known to many beekeepers and is often among the collections of bee plants listed in seedsmen's catalogues. Without doubt it deserves the reputation it enjoys, for it undoubtedly secretes nectar very freely and its bluish-pink flowers are often to be seen covered with bees, the flowers being visited at all hours of the day. It was introduced into Europe from North America towards the end of last century, and there are many references to it in beekeeping literature—English and continental. It has also attracted attention as a fodder plant for livestock, but while satisfactory as green fodder, it is generally considered too fibrous or woody for hay.

Sown in the spring or early summer, flowering commences in about eight weeks, and lasts for four to six weeks, a long period for an annual. Large numbers of flowers are produced by each plant. The plant succeeds in almost any soil and grows up to 2 feet in height. It lacks the showiness associated with a first-class garden plant, but is not out of place as an edging. Grown in quantity by itself it is inclined to lodge. Sown mixed with sweet clover it provides good bee forage in the first season when the clover itself does not flower. It has also been recommended for sowing between the rows of potatoes after the last earthing up, when its growth in no way interferes with the potato crop, and it gives excellent bee forage. In the autumn it is ploughed or dug in as green manure (*Schweizerische Bienen Zeitung*, Jan., 1927). In most areas two crops of flowers may be obtained from the same plot of ground by sowing in late September or early October to stand the winter, and again in early June. The autumn sowing flowers at the end of April and May, while the June sowing provides welcome bee forage in August.

In the phacelia flower the nectar is secreted freely by a disc at the base of the ovary and is protected by special appendages at the base of the stamens. These do not hinder the honey bee. The pollen is pale blue, but of a darker hue in the bees' pollen baskets, and the individual grains are biscuit-shaped, not spherical. Honey obtained from the plant in California, from experimental sowing, is described as light green and of fine flavour (23). Honey has also been obtained from other wild species of *Phacelia* in the southern United States. At Kew the hive bee has been observed working the following species for nectar—*P. viscida*, *P. congesta* and *P. campanularia*. The last mentioned has large bell-shaped blue flowers, and is sometimes to be seen in gardens.

Pheasant's Eye   *Adonis spp.: Ranunculaceae*

The vivid flowers of this popular annual are sometimes visited by bees for pollen.

Pieris   *Pieris floribunda*, *P. japonica: Ericaceae*

These evergreen shrubs, with their white pitcher-shaped flowers, bloom early in the year (March). Bees visit the blossoms, but as the flower-tube is a quarter of an inch or more in length, with a constricted opening, it is doubtful whether the honey bee obtains much nectar unless secretion is very heavy. Conditions for successful cultivation are the same as for rhododendrons.

Plantain   *Plantago lanceolata: Plantaginaceae*

The flowers of this very common weed are available for the greater part of the summer and sometimes attract bees for pollen, but more usually are entirely ignored. The pollen is of the dry wind-borne type which bees do not seem to relish when other pollen (from entomophilous flowers) is available.

The following interesting account of the way bees visit the flowers has been recorded: 'The honey bee flies buzzing to a spike, and while it hovers in the air it spits a little honey on the exserted anthers. Then, still hovering and buzzing, it brushes pollen with the tarsal brushes of its forefeet off the anther, the tone of its humming becoming suddenly higher; in the same instant one sees a cloud of pollen rise from the shaken anthers. After placing the pollen on its hind legs the bee repeats the operation on the same or other spikes, or if it is tired it alights on the spike and creeps upward. . . . ' (11). 'In windy weather the honey bee behaves quite differently when collecting pollen. In these circumstances it flies straight to the spikes, goes once round the zone containing opening flowers, and brushes its legs over the projecting anthers. It is thus able, after the loosely placed pollen has been dispersed by the wind, to obtain still further supplies. Honey bees vary individually in their treatment of these anemophilous flowers.' (16).

Polygonum   *Polygonum spp.: Polygonaceae*

Several wild plants and weeds closely related to buckwheat and belonging to the same genus, *Polygonum*, are good bee plants. These include climbing buckwheat (*P. convolvulus*), a troublesome cornfield weed and the common bistort (*P. bistorta*) with its pretty spikes of small, flesh-coloured flowers. The latter is usually to be found in moist, grassy places. Persicary (*P. persicaria*), a ubiquitous weed not only in Britain but in Europe and America as well, is reputed to be a good nectar plant in some countries. It yields surplus in parts of the United States and Canada, where it comes up freely in cornfields after cultivation has ceased (19), the honey being described as spicy, very dark, and granulating rapidly (3). In Britain this plant does not seem to be very freely worked by bees. The same has been said of it in Germany (11). Other wild British species visited for nectar are water pepper (*P. hydropiper*), frequent in ditches, and the amphibious persicary (*P. amphibium*), a showy aquatic plant. The latter may be useful plants for establishing on or around ponds or waste stretches of water to increase the bee pasturage of a district. Nectar secretion in a water plant is probably less liable to be affected by dry spells than are terrestrial plants, provided always of course the drought is not severe enough to dry up the pool or pond.

Some of the polygonums cultivated in gardens are good bee plants, the most noteworthy being perhaps *Reynoutria sachalinensis* (formerly *P. sachalinense*), introduced from the Sachalin Islands off the coast of Siberia in 1869, an imposing giant perennial with stems 10 to 12 feet high and leaves 1 foot in length. It is ideal for semi-wild places and a good nectar plant with its clusters of greenish-white flowers in late summer (*American Bee Journal*, 1943, 438).

Poplar   *Populus spp.: Salicaceae*

Poplars are sometimes useful for early pollen. The half dozen or so different kinds that are commonly to be seen produce their catkins in March or April. These may be conspicuous on account of their red or purple anthers. In the balsam poplars (e.g. *P. trichocarpa* from western Canada) the buds are covered with a fragrant balsamic gum. Bees have been observed buzzing round the trees and collecting it, presumably for use as propolis.

Poppy   *Papaver spp.: Papaveraceae*

Both wild and garden poppies produce pollen in abundance. This is
generally dark coloured and seems to be much to the bees' liking,
for they visit the flowers in large numbers for it, even when many
other sources of pollen are available. The same applies to bumble
bees. In the large single flowers of the Oriental poppies it is not
unusual to see three or four hive bees together all revelling in the
rich store of pollen.

There seems to be some evidence that a certain amount of nectar
may also be available from poppies at times, particularly the common field poppy, *Papaver rhoeas* (4). Various reports have appeared
to the effect that the poppy may have a narcotizing or stupefying
effect on bees. A German observer noticed that 90 per cent of his
bees returning with pollen loads from the field poppy had difficulty
in finding the entrance of their hive (*Bee World*, 1936, 47). As the
poppy is the source of opium and various narcotic principles this
may not be surprising.

Potentilla   *Potentilla spp.: Rosaceae*

Some of the potentillas or cinquefoils, both wild and cultivated, are
visited by bees for nectar. There are more than a dozen in the British
flora. One of the best known is the silver-weed (*P. anserina*), a
creeping plant with silvery leaves and yellow flowers that occurs
more or less everywhere. Bees visit the flowers at times, but not in
large numbers. The same applies to the shrubby cinquefoil (*P.
fruticosa*), a rare plant in the wild state, but grown in gardens.

Primrose   *Primula spp.: Primulaceae*

There have been reports that bees visit primroses (27), but it is
doubtful whether they are of any value, certainly not for nectar.

Privet   *Ligustrum vulgare: Oleaceae*

The flowers of privet yield nectar freely and this is readily collected
by hive bees. Unfortunately, it produces a strong flavoured, bitter
honey, thick and dark coloured, which will spoil any other honey
with which it is mixed. However, privet in flower is seldom sufficiently abundant for this to occur, and in most districts more a source of
good than evil to the beekeeper. Overgrown or neglected privet
hedges, bearing a profusion of blossom, are a common enough sight.

They are generally viewed with favour by urban beekeepers, especially as they provide food for bees at a time when limes are over and there is little else.

The common privet (*L. vulgare*) is wild and sometimes common in chalk scrub and around beechwoods in the south of England, flowering usually in June and July. As a hedge plant it has been largely superseded by Japanese privet (*L. ovalifolium*) which retains its leaves better. This privet flowers later (August and September) when allowed to do so and not regularly trimmed.

The clusters of white flowers of privet have a heavy penetrating odour, rather objectionable to most people. The flower-tubes are short and the honey bee has no difficulty in getting at the nectar. Sometimes nectar secretion is so copious overnight as to reach half-way up the flower-tube in the morning.

Other privets are sometimes cultivated as ornamental shrubs. Of these, Chinese privet (*L. sinense*) is one of the most handsome, with its large feathery masses of bloom in June and July. These are frequently covered with bees. Even with the limes out at Kew bees continue to work the blossoms.

Puff Ball   *Lycoperdon giganteum: Fungi*

This interesting fungus, which attains the dimensions of a football and is edible in the young stage, is of interest to the beekeeper. It is quite common in some parts of the country. The smoke from it has long been used for stupefying bees, the fungus being first cut into slices and dried in the sun before being burned. There is reference to its being used in this way by 'skeppists' in the older bee literature. It was, and may still be, used in parts of Scotland. In more recent years its use as smoker fuel has been suggested, also the thick fleshy fungi common on tree trunks (*Bee World*, 1936, 6).

Pyracantha   *Pyracantha coccinea: Rosaceae*

The masses of white blossoms of this popular evergreen shrub are attractive to bees when they appear in May or June. Nectar is secreted very freely by the flowers at times. Pollen is also collected. In prolonged drought the nectar flow may cease, for it has been noticed that bees are inclined to pay little attention to the blossoms during such periods.

The pyracantha, or firethorn as it is also called, was introduced to cultivation in Britain from southern Europe over 300 years ago.

There are several varieties of it, and it is widely grown, particularly against walls, for which it is well adapted. Grown this way it produces its attractive fruits more freely than in the open. Its one weakness is that it transplants badly except when young.

Quince   *Cydonia oblonga: Rosaceae*

Like the medlar the quince is one of those hardy fruits that cannot claim wide popularity in Britain and is not so extensively grown as in many other countries. It reached this country from southern Europe, where it is naturalized, but its real origin is uncertain as is the case with many other cultivated plants. The quince has solitary flowers. These are similar in structure to those of the apple and pear but are larger and appear later. They are a source of nectar and pollen to the honey bee. The so-called Japanese quince or japonica is a closely related plant grown for ornamental purposes (see Japonica).

Radish   *Raphanus sativus: Cruciferae*

When left neglected in the vegetable garden or when grown for seed the flowers of the radish attract bees for nectar.

Of greater importance to the beekeeper however is the wild radish (*R. raphanistrum*), also known as white- or jointed-charlock or runch. It is a troublesome weed in cornfields and arable land on all types of soil. It grows 1 to 2 feet high and flowers in June or July with straw-coloured, sometimes white, distinctly-veined flowers. The plant is often confused with ordinary charlock, which it resembles, but may be distinguished by its jointed seed pods. The roots smell like radish but are more pungent. As a bee plant it is probably of about the same value as charlock and useful to the beekeeper when present in quantity. The honey is also believed to be similar and light in colour (14).

Ragged Robin   *Lychnis flos-cuculi: Caryophyllaceae*

Bees are known to visit the rose-coloured flowers of this common wild plant for nectar and pollen. Nectar is secreted at the base of the stamens and is probably difficult for the honey bee to reach on account of the length of the flower-tube.

Ragwort   *Senecio jacobaea: Compositae*

Ragwort is frequently a troublesome weed in Britain as well as in many other countries. It is common in waste places and in meadowland and may be anythng from 1 to 4 feet high. With its masses of yellow flowers, produced any time from June to October it may be a striking feature of the landscape, particularly on light, medium or calcareous soils. In meadowland it is mainly troublesome where cattle only are grazed, for they do not touch the plant, whereas sheep eat out the bud of the rosette of leaves in the early stages and so prevent or retard its further development. The whole plant has a strong smell which accounts for the name stinking Willie applied to it in Scotland.

Ragwort is always attractive to bees and a prolific source of nectar and pollen, particularly late in the season when other sources are over. Unfortunately the honey, like the plant, is strong flavoured, almost bitter in fact, and liable to spoil other honey if present in any quantity. This does not detract from its usefulness for the bees' own consumption and for winter stores. The honey is deep yellow in colour, the aroma being characteristic and strong as well as the flavour. When bees work this plant the wax produced is also a deep yellow, probably stained from oil in the pollen as in sainfoin. The pollen itself is deep yellow in colour. When prolonged drought causes a failure with white clover, ragwort honey may be stored in quantity if the plant is prevalent for it is fairly drought resistant. Much Irish honey is believed to contain a certain amount of ragwort honey (14).

The so-called Oxford ragwort (S. squalidus), a rather similar but more elegant plant that originated from Europe, is now common in some areas. It flowers from June to November and often attracts bees for nectar and pollen.

Raspberry   Rubus ideaus: Rosaceae

Where raspberries are cultivated on a large scale for market they provide valuable bee forage, for the flowers are good nectar yielders, giving a high-quality honey. Furthermore, flowering generally takes place at a most appropriate time—between the blossoming of fruit trees and the first appearance of white clover. Raspberries have long been grown on a field scale in Scotland where they thrive and claim more attention than they do further south, where there is a wider range of fruits available for cultivation. The wild raspberry is also more common in the north than the south, and is the progenitor of the cultivated sorts, which it closely resembles. However, its fruits are smaller and more prone to be dry. The plant may often be seen thriving in places where fires have been. Many of the so-called wild raspberries are the offspring of cultivated kinds through seed being spread by birds.

Raspberry flowers are mostly pendulous and the nectar well protected from rain. Bees are able to work them when those of other plants have been spoiled by rain and visit them even in dull weather. The nectar is easily available to the honey bee and is secreted within the stamen circle at the base of the flower. It first appears as drops and if not removed by insects soon covers the base of the flower. Raspberry flowers are undoubtedly exceptionally attractive to the honey bee. Besides nectar, masses of white pollen are collected from them.

The honey from raspberry is light in colour and of a delicate flavour. Some consider it superior to any other table honey— 'partaking somewhat of the exquisite flavour of the fruit itself . . . while its delicious comb almost melts in the mouth' (20).

The honey bee is unable to puncture the skins of fruits like plums, cherries, and grapes, for they are too tough for its jaws. This does not apply in the case of the raspberry, however, with its very delicate skin. Not infrequently bees collect the sweet juice from ripe and overripe raspberries. This is considered to be the source of the so-called red honey that is reported by British beekeepers from time to time (*Bee World*, 1942, 45).

The dewberry (*R. caesius*) and many other species of *Rubus* and the hybrid berries often cultivated are attractive to bees but are not usually available in any quantity.

Rhododendron *Rhododendron ponticum: Ericaeae*

Rhododendrons do not usually offer much attraction to the honey bee although the flowers are much visited by bumble bees which are better able to get at the nectar with their longer tongues. The flowers are similar in structure to those of azaleas where the same applies. The small-flowered rhododendrons, such as the dwarf kinds often seen in rock gardens, are doubtless better suited to the hive bee.

The common purple-flowered or pontic rhododendron (*R. ponticum*) that is naturalized and occurs so freely throughout the country, often on the outskirts of woods, is known to be worked for nectar by honey bees at times, possibly only when nectar is secreted copiously and can be easily reached. This plant has been known as a source of poisonous honey in southern Europe from classical times and there are at least two instances where honey with harmful or deleterious properties in Britain is suspected of having been obtained from it (see remarks on poisonous honey, Section 1). The pollen of rhododendrons is like that of the heaths, in tetrads, but is larger

Rock Beauty   *Petrocallis pyrenaica: Cruciferae*

This handsome little Alpine plant with its cushions of leaves 2 to 3 inches high is not unlike a saxifrage. It bears pale lilac, sweet-scented flowers in April, which yield nectar and pollen.

Rock Rose   *Helianthemum spp.: Cistaceae*

Both the wild and the garden rock roses, or sun roses as they are sometimes called, may attract bees in large numbers in bright sunny weather—the only time that the flowers are open in fact. The wild kinds are most common on chalk hills and on cliffs and hillsides near the sea. In these situations their yellow flowers may appear in great profusion from May until July. Each flower bears numerous stamens and is a source of much pollen. It is mainly for pollen that bees visit the flowers. This applies also to the garden rock roses with their bright flowers of many shades, which are often favoured for rock gardens and stony slopes.

Rocket   *Hesperis matronalis: Cruciferae*

This favourite old garden plant, with its sweet-scented mauve or white flowers, is sometimes popular with bees for pollen. The single form is easily naturalized in shrubberies and has been found apparently wild in some parts of the country.

Rose   *Rosa spp.: Rosaceae*

The numerous wild roses that beautify the countryside from June onwards yield pollen in abundance and are sometimes visited by honey bees on this account. There has been discussion in the past as to whether roses are not also a source of some nectar but the consensus of opinion is against this, at least in so far as conditions in Britain are concerned. The same applies to garden roses when these are of the single-flowered type. The pollen of the rose is very like that of the apple and is gathered greedily by bees at times.

Rosemary   *Rosmarinus officinalis: Labiatae*

The pale mauve flowers of this popular garden plant are much loved by bees, who crowd over them when they appear from April till June. This evergreen shrub is a native of the Mediterranean region and Asia Minor and is not hardy everywhere in Britain. It succeeds best in warm situations on light dry soils. On chalk the bush grows smaller but is more fragrant.

The famous Narbonne honey, which in normal times may be purchased from some of the big London stores, is derived largely from this plant. Rosemary is an important source of honey in many parts of Spain and imparts something of its own fragrance to the honey gathered from it. The shrub is well suited as a low evergreen hedge, particularly when one is required for the bee garden, for it never fails to attract.

Rue   *Ruta graveolens: Rutaceae*

This is of little account as a bee plant although bees are known to visit the greenish-yellow flowers for nectar (1). In them the nectar is exposed and so attracts numerous short-tongued insects as well. Rue was more generally grown for medicinal purposes in bygone days.

Several species of meadow-rue (*Thalictrum*) are grown in gardens but more for their attractive foliage than for their flowers, which have no special appeal. However, the honey bee considers them attractive for pollen and works them assiduously for it.

Safflower   *Carthamus tinctorius: Compositae*

Safflower has long been an important oil seed crop in India and the East and has been considered as an alternative crop to the sunflower in other countries. There are many varieties ranging in height to 5 feet. The thistle-like flowers are sometimes used as a dye. They secrete nectar very freely and are much visited by bees. The plant will grow and flower in the south of England but for profitable seed production doubtless needs a warmer climate.

Sage   *Salvia officinalis: Labiatae*

This culinary herb, so much used for flavouring, has been cultivated in Britain for many centuries and originated in the Mediterranean region. In its native haunts it is sometimes the most common plant of the low shrubby vegetation so typical of the hillsides in that area. It is there the source of much fine honey, light in colour, of good flavour and slow to granulate.

Sage does not grow quite so luxuriantly in the cool, moist English climate, nevertheless apart from its cultivation in gardens for home use, it is grown on a field scale in some parts of the country, notably the market garden districts of Kent, Surrey, Cambridgeshire, Bedfordshire and Worcestershire. It is not fastidious in regard to soil. It may be harvested periodically throughout the summer or at the end of the season and dried, which is more desirable from the bee-keeper's point of view as the flowers afford first-class bee forage. Two sorts of sage are commonly grown—the broad and the narrow leaved. Red or purple sage is sometimes cultivated, but mainly for decorative or medicinal purposes.

Two wild sages, meadow sage (*S. pratensis*) and clary sage (*S. verbenaca*), are good bee plants but probably nowhere sufficiently abundant to be a source of honey. Many varieties of the former are grown in flower gardens. Clary sage is also cultivated and is reputed to be a useful bee plant in parts of Australia where it has become naturalized (19). In California some of the wild sages are first-class nectar plants and yield a high-grade honey which is very slow to granulate.

A sage closely allied to culinary sage and known as Balkan sage is sometimes to be seen in collections of bee plants. This was made known to British beekeepers by Mr Herbert Mace who states in regard to it: 'During my war service in the Balkans, I collected seed from thirty plants new to me and attractive to bees. On my return these were sown with varying results. Some failed to germinate, others did not grow well in this climate, or failed to seed, and only one proved entirely satisfactory. This is a species of sage, hitherto unknown to English gardeners, but said to be near *Salvia amplexi-caulis*. It is a great acquisition, both as a border and bee plant. It is a perennial and two plants grown from the original seed are still vigorous after fifteen years, but it does not spread or become weedy. It attains a height of 2 to 4 feet, forming a dense bush bearing a mass of glorious blue flower amongst purple bracts, making it a most striking object. It begins to bloom in June and continues till October if faded sprays are cut off. In fine or dull weather the plants are besiged by bees and butterflies ...' (*The Beekeeping Annual List*, 1936).

Salvia   *Salvia spp.: Labiatae*

Some of the cultivated or garden salvias are attractive to bees, but in others the flower-tube is too long for the honey bee to be able to reach the nectar. This applies to the commonest of all salvias, the scarlet salvia (*S. splendens*), so much used for bedding in English gardens, but not hardy.

The best salvia for bees in the writer's experience is *S. superba* (formerly called *S. virgata nemorosa*) which is a hybrid and originated from eastern Europe. It is a perennial and perfectly hardy, reaching 2 to 3 feet in height, with erect growth, requiring no staking. Flowering commences in June, when the numerous spikes of purple-blue flowers are a pretty sight. It is in full bloom for a month to six weeks and continues to flower until September. During the whole of this time the flowers swarm with honey bees working for nectar, which they have no difficulty in reaching. A large circular bed of this plant at Kew is often so covered with bees in June that it almost looks as though a swarm were alighting, or in the vicinity. Besides being a good bee plant *S. superba* is also a first-class garden plant with its handsome flowers, long flowering season, and ease of cultivation. It needs full sun like most salvias, and is propagated by division of the roots or cuttings, not seeds. It does not form viable seed and the pollen also is abortive which is not unusual in hybrid plants. No bee garden should be without this plant.

Saponaria   *Saponaria spp.: Caryophyllaceae*

These garden annuals are sometimes visited by bees, probably mainly for pollen, those sown in autumn for early spring blooming being the most useful in this respect.

Sassafras   *Sassafras albidum* (formerly *S. officinale*): *Lauraceae*

This North American tree is occasionally seen in cultivation in the warmer parts of the British Isles, and is of interest on account of its aromatic nature—being the source of oil of sassafras. Its small greenish-yellow flowers, which appear in May for about two weeks, are visited by honey bees for nectar.

Savory   *Satureia montana, S. hortensis: Labiatae*

Savory is one of the less common of the culinary herbs in English gardens. With its strong flavour it is sometimes used for seasoning, like thyme, and in France is cooked with broad beans as mint is cooked with peas. It belongs to the mint family. There are two sorts, summer savory (*S. hortensis*) an annual, and winter savory (*S. montana*) a perennial. Both are good bee plants and well worked for

G

nectar, especially the latter. They flower in June and July and have pale rather insignificant flowers.

Saxifrage   *Saxifraga spp.: Saxifragaceae*

The saxifrages, wild or cultivated, are not in the front rank among bee plants, but honey bees frequently visit the flowers. In them the nectar is very exposed and only secreted in sunny weather when flies and other short-tongued insects immediately have access to it. Such flowers are not usually great favourites with honey bees. Perhaps they do not care to rub shoulders with flies or drink from the same vessel! The same applies in the large family *Umbelliferae*, to which the carrot belongs.

Among garden saxifrages London pride (*S. urbium*) is best known. It possesses a beautifully-marked pollen grain, a delicate flesh colour in honey (27).

Scabious   *Scabiosa spp.: Dipsaceae*

Wild and garden scabious are good nectar plants. The wild species are all freely visited by the honey bee, especially the field scabious (*Knautia arvensis*) and devil's bit (*Succisa pratensis*), and are very common in some parts of the country. All have handsome blue or lilac flowers and are in bloom mainly in July and August. One writer records much nectar being gathered from wild scabious in September (*Bee Craft*, 1938, 313). The secretion of nectar in the flower is of interest for it takes place on the upper surface of the ovary and is protected from rain by hairs in the corolla tube. In the field scabious this varies from 4–9 mm. according to the position of the flower in the flower-head, and the nectar is generally easily reached by the honey bee. The pale yellow pollen is also collected.

Over a dozen of the wild scabious from other lands in cultivation at Kew have been observed being freely worked for nectar. Popularity with bees applies also to garden kinds with their flowers of so many different shades. The so-called sheep's bit scabious (*Jasione montana: Campanulaceae*) is an entirely different plant, but is also common in many districts and visited by hive bees for nectar.

Schizopetalon   *Schizopetalon walkeri: Cruciferae*

The unusual and scented white flowers of this half hardy annual from Chile are attractive to bees—like most crucifers. The plant is not often grown in gardens.

Scurvy-grass   *Cochlearia officinalis: Cruciferae*

Honey bees work the white flowers of ths old-time medicinal plant
freely at times. It is a characteristic plant of the seashore, blossoming
in May.

Sea-cabbage   *Brassica oleracea: Cruciferae*

The sea-cabbage is often a conspicuous plant on sea cliffs and is
known to be the parent of the cultivated cabbage. Its yellow flowers
yield nectar freely and attract bees like those of its garden relative
(see Brassica).

Sea-holly   *Eryngium spp.: Umbelliferae*

The sea-hollies or eryngiums are often mistaken for thistles which
they closely resemble with their prickly holly-like leaves. Besides the
wild sea-holly (*E. maritimum*) and field eryngo (*E. campestre*), a rare
plant, the flowers of several cultivated or garden kinds are visited by
honey bees for nectar. One of the most interesting of these to the
beekeeper is perhaps *E. giganteum*, from the Caucasus, which bears a
very large inflorescence. This perennial is favoured by bees. It has
been suggested for sowing in waste ground as bee pasturage in
Europe (*Bee World*, 1926, 117), the argument being put forward that
it is unlikely to become a pest because slight injury to the root kills
the plant.

Sea-kale   *Crambe maritima: Cruciferae*

The wild sea-kale is only to be found near the seashore—particularly
in the west of England. Its white flowers appear in June and are
similar to those of the sea-kale of the vegetable garden when this is
allowed to run to seed. Bees visit them for nectar.

Sea-lavender   *Limonium (Statice) spp.: Plumbaginaceae*

Sea-lavender is a well-known plant of the muddy shores and salt
marshes of England. The common species (*L. vulgare*) sometimes
covers wide expanses, producing its flowering stalks, 2 to 3 feet high,
from July to September. The flat-topped clusters of small blossoms,
like lavender in colour, but not in scent for they are odourless, make
it conspicuous from a distance. There are reports of bees working
the blossoms well for nectar and that the honey is light and of good
quality (14). Frequently the plant is most prevalent in remote places,

but where bees have access to it in quantity it is doubtless a useful late source of nectar, for it is available when most other sources, except heather, are over.

Sedges   *Cyperaceae*

Some sedges and rushes have been observed to be worked by honey bees for pollen even when other sources are available (*Bee World*, 1939, 12). The same is known to apply in the case of certain grasses (see Grasses). However, on the whole they are of little consequence to the honey bee. One observer (27) records having seen bees eagerly visiting the flowers of the wood rush (*Luzula campestris*) for pollen.

Self-heal   *Prunella vulgaris: Labiatae*

The purple flowers of this extremely common wild plant are visited by bees on occasions, but it could hardly be called a good bee plant. As the flower-tube is about 8 mm. in length it may be that nectar is not available to the hive bee unless secretion is very abundant.

Senecio   *Senecio spp.: Compositae*

Several senecios attract bees for nectar, notable among them being *S. fluviatilis* (formerly *S. sarracenicus*), sometimes called broad-leaved ragwort. This is not a native plant, but it is naturalized in some areas. It is 4 to 5 feet high and best suited for the semi-wild garden, being rather coarse. The yellow flowers appear from June onwards. The plant is reputed to have been introduced to Britain by the Crusaders.

Serradella   *Ornithopus sativa: Leguminosae*

Serradella is often cultivated in European countries as a fodder plant, but seldom in Britain and only in the south. It is an annual, clover-like plant, 1 to 2 feet high and well suited for poor, sandy soils. The small pinkish flowers may be available for as long as three months, and are considered to be a good source of nectar. The plant does not ripen its seeds evenly or well in Britain and is really happier in a warmer climate.

Sida   *Sida hermaphrodita* (formerly *S. nabaca*): *Malvaceae*

The white flowers of this vigorous, mallow-like perennial—not often cultivated—are visited for nectar and pollen by bees.

Sidalcea   *Sidalcea malvaeflora: Malvaceae*

The sidalceas of the flower border are visited for nectar and pollen, mainly the latter. These mallow-like perennials with their rose-purple or white flowers, thrive in almost any soil. Known as wild hollyhock they are of importance for nectar and pollen in California, where they are wild and are often to be seen along ditches and roadsides (23).

Silphium   *Silphium spp.: Compositae*

These tall, hardy perennials, which resemble sunflowers, are not often seen in the flower garden, being somewhat coarse. They are vigorous and able to thrive in the heaviest clay soils and are suited for the semi-wild garden. Several species have been observed being worked for nectar and pollen at Kew. *S. perfoliatum* (cup plant) is reputed to be abundant and afford good bee forage in parts of the Mississippi valley (19).

Skimmia   *Skimmia japonica: Rutaceae*

Both male and female flowers of this useful and much-cultivated evergreen shrub secrete nectar freely but nevertheless do not seem to attract the honey bee to any extent. The flowers appear early in the year and are followed by bright red berries—one of the main attractions of this Japanese shrub. Bees have been observed collecting pollen from the male flowers in April at Kew

Snowberry   *Symphoricarpos albus: Caprifoliaceae*

This hardy North American shrub is common throughout the country and is often seen in shrubberies, as a rough hedge plant, or apparently wild, for it is extensively naturalized. As it suckers freely it often forms dense thickets. It is easily distinguished from all other shrubs by its pure white fruits, which account for its name.

The small, bell-shaped, pinkish-white flowers appear from June to August. They are not conspicuous, but must be great nectar yielders judging by the way they attract hive bees, bumble bees and wasps. There is not a large number of flowers out at any one time, but the flowering period is a long one and the flowers may be worked all day. On warm summer evenings honey bees have been observed on the flowers at quite a late hour. The shrub has been recommended for planting to improve the August honey flow where waste or suitable land exists (*Bee World*, 1936, 20) for, once established, it is well able to look after itself. In its native land bees are said to work it in preference to white clover and excellent honey is obtained from it (20).

Other species of *Symphoricarpos*, sometimes grown for their ornamental fruits, are known to be first-class nectar plants, such as *S. occidentalis* (wolf berry) and *S. orbiculatus* (Indian currant or coral berry) (20).

Snowdrop   *Galanthus nivalis: Amaryllidaceae*

Always one of the best loved of early spring flowers, the snowdrop is often to be found in bloom in woods and fields as early as February. It remains in bloom for several weeks. When weather is suitable, bees seek the flowers with zest and make good use of the bright yellow pollen which it furnishes. Where the plants grow abundantly bees sometimes swarm over the flowers on bright sunny days. Besides yielding pollen the flowers secrete a certain amount of nectar. This is formed in depressions on the inner sides of the petals and at the base of the flowers. It is not always easily seen except when flowers have been kept in a warm room overnight.

In collecting pollen from the snowdrop the honey bee inserts its head, front and middle legs into the flower, clinging with its hind legs to a petal. It then brushes the anthers with the fore and middle legs and deposits the pollen obtained in the pollen baskets of the hind legs (16).

Snowdrop Tree   *Halesia carolina: Styracaceae*

The snowdrop or silver bell tree, a native of the south-eastern United States, is sometimes seen in English gardens. It produces clusters of pendulous white flowers, not unlike snowdrops, in May. These are worked fairly freely by bees in some seasons.

Snowflake   *Leucojum spp.: Amaryllidaceae*

The spring snowflake (*L. vernum*) like its close relative the snowdrop, is a useful early source of pollen for the honey bee and may also yield a little nectar given favourable weather. It is usually in flower in March and April, the fragrant drooping flowers resembling large snowdrops. The plant is very local in its distribution. The summer snowflake (*L. aestivum*) is similar, but is a larger plant and flowers later.

Solomon's Seal   *Polygonatum spp.: Liliaceae*

There are three species of Solomon's seal native to Britain but all are
rare or local in distribution. One of these, the so-called angular
Solomon's seal (*P. officinale*) which flowers in May and June, is
reported to be an important nectar plant in parts of Holland (*Bee
World*, 1937, 48).

Sow-thistle   *Sonchus spp.: Compositae*

The flowers of the corn or perennial sow-thistle (*S. arvensis*) and the
common annual sow-thistle (*S. oleraceus*) are visited by bees. Both
are weeds up and down the country, the former mainly in cornfields
and the latter in waste places, along roadsides, etc. The corn sow-
thistle with its creeping roots is often a pestilential weed in fields and
flowers from August onwards, its yellow flower-heads, 40 to 50 mm.
across, attracting bees for nectar and pollen. In the Prairie Provinces
of Canada this European weed is now equally common and is
regarded as a good honey plant (*Bee World*, 1923, 4). The common
sow-thistle in Britain is mainly visited for pollen: the corolla tube
being 8 to 12 mm. in length.

Spurge   *Euphorbia spp.: Euphorbiaceae*

The greenish-yellow flowers of the spurges (particularly *E. esula*) are
sometimes visited by bees for nectar, but on the whole do not offer
much attraction and are generally neglected when other plants are
available. Bees collecting nectar and pollen from spurges at Kew have
been observed with dark brown pollen loads.

Spurrey   *Spergula arvensis: Caryophyllaceae*

Although primarily a weed of light sandy soils, especially those
deficient in lime, corn spurrey has been grown as a forage crop for
sheep in some countries. Its small white flowers are produced from
June till August. These are closed in unfavourable weather and only
open fully in sunlight. Nectar may be formed at the base of the
stamens, and honey bees are known to visit the flowers.

Stachys   *Stachys spp.: Labiatae*

The genus *Stachys* is a large one with over 200 species. About half a
dozen (commonly called wound-worts) are native to the British Isles.
Others are sometimes cultivated as ornamental plants. In those cases

where the flower-tube is not too long the honey bee may be a frequent visitor for nectar. One species (*S. recta*, with yellow flowers) is reputed to be abundant in Czechoslovakia and to be an important source of honey there, the honey being water white, dense and slow to granulate (27).

St. John's Wort   *Hypericum spp.: Hypericaceae*

The St John's wort of the flower garden, as well as the many wild kinds, yields pollen in abundance, each flower possessing numerous stamens. This is frequently collected by bees, and when packed in their pollen baskets is orange in colour. The flowers appear to yield no nectar whatsoever.

Stonecrop   *Sedum spp.: Crassulaceae*

The stonecrops closely resemble the saxifrages and occur in similar situations, being often seen on old walls and rocky banks in the wild state. The flowers attract bees to some extent, especially those of the large purple-flowered wild sedum (*S. telephium*), of which there are many garden forms or varieties. The showy, rose-purple flowers of the much-grown Japanese sedum (*S. spectabile*) which generally appear in mid August also claim the attention of hive bees. This accommodating plant will grow in any soil, or in shade, and withstands great extremes of heat or cold well.

Strawberry   *Fragaria vesca: Rosaceae*

The strawberry has often been overrated as a bee plant. Bees do visit the blossoms, sometimes very freely, but mainly for pollen. This is one of the instances where the hive bee is of great service to the fruit grower for some varieties of strawberry are self-sterile or deficient in pollen and cross pollination by insect agency is of primary importance.

Strawberries are extensively grown for market and on a field scale in many parts of the country, but never does one hear of a beekeeper obtaining surplus from this source. Actually a nectary is present in the flower and takes the form of a narrow fleshy ring on the receptacle or the base of the flower. Nectar is secreted at times but not in large amounts. It is concealed and partly protected by the carpels and stamens. The pollen is of a pale yellow colour.

The flowers of wild strawberries are similar in structure but smaller and probably of about equal value to the honey bee. Wild strawberries are usually to be seen in flower in May and June, but are not so prevalent in Britain as in some parts of Europe.

Strawberry Tree   *Arbutus unedo: Ericaceae*

This rather unusual evergreen tree is sometimes to be seen in cultivation, but in the British Isles occurs wild only in southern Ireland. The pinkish-white, pitcher-shaped flowers are produced at an unusual time, from October to December, and develop into ripe fruits the following autumn. Bees have been observed visiting the flowers for nectar and pollen on fine autumn days. Honey is obtained from the strawberry tree in some parts of the Mediterranean region, e.g. Sardinia and Greece, and is described as lemon yellow in colour with an aromatic odour and bitter to the taste (30).

A similar tree (*A. menziesii*) from North America is also sometimes cultivated. In its native land it is a source of good-quality honey (23).

Sumac   *Rhus spp.: Anacardiaceae*

There are about 150 sumacs or species of *Rhus* distributed throughout the world. None is native to Britain but a few are sometimes cultivated for their ornamental foliage. The flowers are generally small, inconspicuous and dull in colour but often very rich in nectar. Several of the North American species are sources of honey. This is inclined to be strong flavoured at first (19). Two are frequently to be seen in English gardens, viz. stagshorn sumac (*R. typhina*) a shrub or small flat-topped tree, and poison ivy (*R. toxicodendron*) a creeper, both grown for their fine-coloured autumn foliage. At Kew a Chinese sumac (*R. potaninii*), a small tree, is covered with blossom in late May or June and bees literally swarm over it.

Sunflower   *Helianthus spp.: Compositae*

The many different kinds of garden sunflower, whether annual or perennial, tall or dwarf, are all more or less good bee plants— double forms excepted. The tall perennial kinds are mostly autumn flowering and may be numbered among the useful, although minor, late sources of nectar and pollen available to urban beekeepers. Some are very late in flowering and provided the flowers are not damaged by frost, are very freely worked, especially for pollen, on fine autumn days. This is a time when there is little else available. In their native land (North America) they are sometimes prevalent enough in the wild state to yield some surplus honey which is described as amber or dark with a characteristic flavour not disliked by most people (20). The large flower-heads of the annual giant sunflower (*H. annuus*) which is grown for its seed for use as poultry food or for oil, are also visited by honey bees for nectar or pollen, along with numerous other insects. This is an important field crop in some countries and honey is obtained from it. According to a French writer (*L'Apiculture Française*, April, 1932) the leaves and stems, when cut green and left to dry, form excellent smoker fuel, giving a dense smoke and burning slowly. The seeds are liked by tits and may help in keeping them out of mischief in an apiary. A gummy secretion sometimes present on the flower-heads may also claim the attention of bees, possibly, as a source of propolis.

Swedish Whitebeam   *Sorbus intermedia: Rosaceae*

This large tree resembling the whitebeam (*S. aria*) and sometimes planted in parks and gardens is common in northern and central Europe, but occurs also in some places in the west of England. It produces masses of white flowers in early June at Kew and these are very freely worked for nectar.

Tamarisk   *Tamarix spp.: Tamaricaceae*

These are favourite shrubs for gardens near the sea, being very well suited for exposed seaside conditions. They are either grown for ornament on account of their feathery foliage and pretty flowers or as hedges and wind-breaks. The common tamarisk (*T. gallica* or *T. anglica*) is much used in this way, being easily grown from cuttings planted direct. It has become freely naturalized in some seaside districts. Flowering takes place in July and August and the bunches of small pink flowers attract bees for nectar. The same plant occurs in other countries where it is sometimes a source of surplus honey, as well as being useful for pollen (23; 4). The honey, however, is reputed to be of poor quality and liable to spoil better-quality honey if mixed with it (*American Bee Journal*, March, 1941).

Some of the garden forms of tamarisk (*T. tetrandra*) are in flower earlier and also attract bees.

Tansy   *Chrysanthemum vulgare: Compositae*

Honey bees do visit the yellow flower-heads of this very common wild plant and have been observed probing for nectar as well as gathering pollen. However, they do not appear to visit it in large numbers. Possibly the strong smell does not appeal to them. It can hardly be called a good bee plant.

The dried flower-heads, leaves and stalks are said to make good smoker fuel, burning slowly and with a pleasant aroma (*Bee World*, 1935, 124).

Tea Tree   *Lycium chinense: Solanaceae*

The writer has never observed bees taking much notice of the purple flowers of this common plant, known also as Chinese box thorn, although some observers have recorded its being visited by hive bees for nectar and pollen (*Bee Craft*, 1936). With its long, arching, somewhat spiny branches and orange berries it is often to be seen wild in hedges and is a familiar sight near the sea.

Teasel   *Dipsacus fullonum: Dipsaceae*

The spiny seed-head of the fuller's teasel is still extensively used in cloth manufacture for raising the nap of woollen cloth. Most of the seed-heads have been imported from European countries in the past but the crop has been grown in many parts of Britain. Teasel is a biennial and flowers in its second season when the mauve flowers in the thistle-like flower-heads are a great attraction to the honey bee and yeld nectar freely. Bees visit them all day long. Flowering commences in July in most districts and as only a few flowers open at a time, starting at the bottom of the flower-head, they are available for several weeks. Honey has been obtained from this plant in countries where it is more extensively grown, being described as thin but light in colour and of good flavour (19).

The wild teasel (*D. sylvestris*), which is similar except for the spines on the seed-head being straight and not curved, is common in many areas and is often to be seen in waste places and along water courses. The flowers appear to be equally attractive to bees. A European species (*D. laciniatus*) with deeply-cut leaves, which is sometimes grown for ornament, also attracts bees when in flower at Kew.

Thistles   *Compositae*

A large number of plants fall into the general category of thistles. Although many of them are troublesome weeds they may be good bee plants and supply nectar and pollen in abundance.

The most important thistle from the beekeeper's point of view is the ubiquitous field thistle (*Cirsium arvense*), sometimes called Canada thistle or creeping thistle on account of its ability to spread by means of its creeping roots. This it may do very quickly and accounts for the difficulty in eradicating it, for the smallest piece left in the soil is capable of forming a new plant. It is also spread by means of seeds. It is certainly one of the worst weeds the farmer and gardener has to contend with, not only in the British Isles but in many other lands. It is common in and around cultivated fields and in pastures, especially those grazed only by sheep. It sometimes takes almost complete possession of abandoned fields and waste land.

The field thistle grows from 1 to 4 feet high according to soil conditions and is in flower from July onwards, producing rose-purple flower-heads each with 100 or more individual flowers or florets. These secrete nectar abundantly. The flower-tube is 8 to 12 mm. long and terminates in a short bell. Nectar quite often rises in the tube as high as this bell and so is easily available to the honey bee as well as other insects. The pollen, which is sticky, is also greedily collected by bees. The individual grain is spherical and spiny when seen under the microscope, as is common in this family. It is often found in honey.

Honey from the field thistle is of good quality, light in colour and of excellent flavour, comparing well with lime and clover (20). In parts of Canada (Ontario) this honey is often obtained but not in Britain where its main value lies in assisting to build up the bees' winter stores in the latter part of the season.

Some thistles have so long a flower-tube that the nectar is out of reach of the honey bee or only rarely available. However, there are several thistles that afford useful late-season bee forage. Hive bees have been observed visiting the flowers of the following: spear thistle (*C. vulgare*, formerly *C. lanceolatum*), melancholy thistle (*C. heterophyllum*), marsh thistle (*C. palustre*), welted thistle (*Carduus acanthoides*, formerly *C. crispus*), musk thistle (*Carduus nutans*) and cotton thistle (*Onopordon acanthium*).

Thrift   *Armeria maritima* (formerly *Statice maritima*):
                                                        Plumbaginaceae

The pink flowers of the common thrift or sea pink are a familiar sight near the seashore in June and July. In salt marshes they are sometimes extremely abundant, the plants growing everywhere in dense cushion-like tufts. They often grow in association with sea-

lavender, a closely related plant, which, however, blooms later.
Thrift occurs also in mountainous areas. It is considered to be a good
source of nectar when bees have access to it and useful for supple-
menting the nectar obtained from other sources. Surplus honey has
been obtained from it (9).

Besides occurring wild, thrift is commonly grown as a garden
plant, particularly in coastal districts. There are many varieties and
introduced forms in cultivation which are also attractive to bees.

Thyme   *Thymus spp.: Labiatae*

Both wild thyme and the garden thyme used for seasoning are
first-rate bee plants and are well worked for nectar. So also are
lemon thyme and the many varieties with pink or mauve flowers or
variegated leaves grown in the flower garden or rock garden.

Wild thyme (*T. serpyllum*) is very abundant in some parts of the
country, especially in the chalk districts, on moorland, and on dry
pastures. It may even impart a purplish colour to the landscape when
in full flower in June and July in spite of its lowly growth. Frequently
it grows in combination with wild marjoram, another good bee plant,
and the two form a useful combination for any nearby beekeeper.
Honey is not obtained pure from wild thyme in Britain as there are
invariably other nectar sources available at the same time. However,
the honey from thyme is known to be of excellent quality and flavour
and its presence in other honey is always likely to improve it. The
honey from Mount Hymettus, famous for its excellence from classical
times, is derived from the wild thyme of that region. The flowers of
thyme are very fragrant and have much the same structure as those of
wild marjoram. Nectar is produced freely and even this has a spicy
flavour.

Garden thyme (*T. vulgaris*) is a native of the Mediterranean
region, where it may cover extensive areas of stony ground and be a
wonderful sight when in flower in spring. It is then valuable for honey.
In Britain thyme is often winter killed, especially on the heavier soils.
Besides being grown for home use it is cultivated on a commercial or
field scale in some districts, notably in Kent, Surrey, Worcestershire,
Bedfordshire and Middlesex. It is either sold as a green herb or used
for drying, being an important constituent of the mixed herbs sold in
packets. While the cultivation of thyme must afford useful supplementary bee fodder for nearby beekeepers it is probably nowhere
sufficiently extensive in Britain for surplus honey.

The ornamental thymes, so well suited for rock gardens or covering dry banks and able to exist in poor soils, are also very attractive to bees. No bee garden should be without some of them. One of the best for attracting bees at Kew, on poor, light sandy soil, is *T. serpyllum coccineus*, which produces an abundance of pretty reddish flowers and soon spreads.

Toad-flax   *Linaria vulgaris: Scrophulariaceae*

In the yellow toad-flax, a common plant along hedgerows and the margins of fields, we have an interesting bee plant. The long spur of the flower is 10 to 12 mm. long and nectar collects in this, frequently to a depth of 5 to 6 mm. or more. Only long-tongued insects are able to take all the nectar, but the hive bee is able to have her share and frequently does when nectar is being produced freely. This she does by creeping right into the flower and sucking out the nectar as far as her tongue will reach, emerging covered with pollen. In view of its prevalence in some districts the yellow toad-flax may be of greater importance as a minor nectar plant in the late summer than is generally supposed. Some of the other wild or cultivated species of *Linaria* are also visited by honey bees (27).

Tobacco   *Nicotiana tabacum: Solanaceae*

Tobacco for smoking has been grown in England and Ireland but in strictly limited quantity and under control. As the plants are normally topped before flowering there are no flowers available for bees. However, in other countries where tobacco is grown and flowering of the plants takes place, there are records of surplus honey being obtained, this being described as strong and dark, rather like that of buckwheat (19).

The flowering tobacco grown in the flower garden is similar. Frequently the flower-tubes are punctured at the base.

Tomato   *Lycopersicum esculentum: Solanaceae*

Tomato flowers are without nectar but bees have been observed visiting them for pollen. However, they seem to offer little attraction when other pollen sources are available.

The strong smell of the tomato plant seems to be offensive to bees and it is said that if one manipulates a stock soon after disbudding tomato plants and with the hands smelling of the juice or oil of the plant bees are prone to sting.

Tree-of-heaven   *Ailanthus altissima: Simaroubaceae*

This large handsome tree from China is often to be seen in cultivation in the south of England. It produces terminal bunches of small white flowers in July or August which bees visit for nectar. The flowers are of two kinds, male and female, the former rather evil smelling with an odour not unlike that of elder flowers—what some might describe as a cat-like odour.

Honey obtained by a London beekeeper (Mr A. Chesnikov) with hives not far from Kensington Gardens was believed, from pollen analysis, to be mainly from this source, the Tree-of-heaven being not uncommon in that neighbourhood as a street tree. This honey, when fresh, was described as possessing an unpleasant aftertaste recalling elder, but after being kept for some time the flavour changed to one of a pleasant muscatel flavour. The honey was of a pale greenish-brown colour and crystallized after about three months with a fine grain ('Ailanthus, Source of a Peculiar London Honey', R. Melville, *Nature*, 1944, 640).

In parts of Europe and North America the Tree-of-heaven has become sufficiently abundant to become a source of surplus honey. It spreads readily by means of seeds and suckers and soon becomes naturalized. In and around Vienna it is plentiful and the honey harvested from it there is said to exceed that from all other sources. This also has been described as greenish in colour with a strong unpleasant flavour at first which disappears in time. In the vicinity of Paris the tree is common but the honey has not a good reputation (4). In the United States the Tree-of-heaven has been extensively planted as a shade tree and has the reputation of being a good nectar yielder. There seems some doubt as to whether it is really a source of ill-tasting honey there (*American Bee Journal*, 1938, 420).

Trillium   *Trillium grandiflorum: Liliaceae*

The white or rose-coloured flowers of the American wood lily sometimes grown in gardens attract bees, but this plant is never seen growing in any quantity.

Tulip   *Tulipa spp.: Liliaceae*

Some observers (27) record having seen bees visiting tulip flowers in numbers for pollen but this does not seem to be general, particularly with the late-flowering kinds. Possibly other kinds of pollen, when available, are preferred by the honey bee. The pollen of many garden tulips is of a dark purplish colour.

Tulip-tree   *Liriodendron tulipifera: Magnoliaceae*

The tulip-tree is often to be seen in cultivation in Britain, where it is quite hardy. Its large tulip-like flowers, that appear in June and July, serve to distinguish it readily from other trees. These attract bees to some extent in hot sunny weather but at other times bees pay no attention to them and no nectar is visible in the flowers.

In its homeland (eastern North America) the tulip-tree is an importance source of honey and equals the lime or basswood as a nectar producer. The honey from it is of good quality, reddish amber in colour and rather strong (30). There it is found to give the best flow when the flowers open late and the weather is warm and dry (20) and that in the more northerly limits of its distribution it is not much use for nectar (*Beekeeping in the Tulip-Tree Region*, U.S. Farmer's Bulletin No. 1222). The climate of Britain probably corresponds more to these northerly limits and the tree is unable to be at its best as a source of nectar.

Flowers of the tulip-tree placed in water overnight in a warm room often show large drops of nectar on the yellow mark at the base of each of the large petals. According to American observers each flower is capable of yielding 'not far from a spoonful of nectar' (20). As much as three grammes of nectar have been collected from a single blossom in America. The nectar is only secreted for a short time by each flower—about a day and a half. Nevertheless, it has been calculated that a single tree may yield over 9 lb. of nectar, equivalent to 2 to $2\frac{1}{2}$ lb. of honey (30).

A closely allied tree, the Chinese tulip-tree (*L. chinense*) is comparatively a newcomer to Britain. The writer has not yet been able to gauge its value as a bee plant or nectar producer.

Tupelo   *Nyssa sylvatica: Nyssaceae*

This tree, also from eastern North America, is famous as a honey producer in its native land and grows quite well in Britain but is not often cultivated. When in flower at Kew it appears to offer no attraction to bees and nectar has not been observed in the flowers. In the wild state it is found chiefly in swamps or moist situations. Possibly with this tree a light dry soil or else the climate of Britain does not favour nectar secretion.

Turnip   *Brassica spp.: Cruciferae*

The turnip is the main root crop of the British farmer and is very
extensively grown, particularly in Ireland, Scotland and the north of
England, where the cooler conditions are well suited to it. The swede
or Swedish turnip is included with it.

Normally turnips are harvested at the end of the first season's
growth, before flowering takes place. Sometimes, however, a certain
amount of flowering occurs in the first year due to various causes—
bolting, late frosts, etc. When grown for seed and left in the ground
for a second season flowering is very profuse and a field of this sort
becomes literally a beekeeper's paradise. The turnip is the equal of
other well-known brassicas as a nectar yielder and produces a similar
type of honey (see Brassica, Mustard, Charlock, etc.).

Valerian   *Valeriana officinalis: Valerianaceae*

The small, pale pink flowers of the wild valerian are commonly
visited by bees for nectar. The plant is often to be found in damp,
shady places near streams and should not be confused with the red
valerian (*Kentranthus ruber*) of the flower garden which is naturalized
in many areas, for this plant has too long a flower-tube to be of any
use to the hive bee.

Valerian grows 3 to 4 feet in height and produces its dense clusters
of fragrant flowers in July and August. It is cultivated in some parts
of the country, especially Derbyshire, for its roots, which are used
medicinally. Unfortunately for the beekeeper the common practice
among growers is to cut off the flowering stalks as soon as they
appear to promote the formation of basal leaves and consequently
larger roots. The honey bee also visits the flowers of the marsh
valerian (*V. dioica*), a much smaller plant, for nectar.

Vegetable Marrow   *Cucurbita pepo: Cucurbitaceae*

The vegetable marrow is the only one of the large group of edible
gourds that may be said to be commonly grown out of doors in the
British Isles. Pumpkins and squashes are occasionally to be seen in
private gardens or allotments, but their cultivation is not general.

The large unisexual flowers of these plants are much visited by
bees as well as by many other insects and supply both nectar and
pollen. Nectar is produced abundantly at times, being secreted at the
bottom of the cup of the flower formed by the fusion of the calyx

H

and the corolla. Honey has been obtained from pumpkins, squashes and melons in countries where they are grown on a large scale (19).

The pollen of the vegetable marrow, like that of other cucurbits, is very adhesive, due to the presence of a thin layer of oily or gummy matter on the grain. Viewed microscopically the individual grain is spherical and prickly and very large (about 150 microns in diameter). It is among the largest found in honey. Both hive and bumble bees seem very partial to the pollen and collect it in large quantities.

Verbascum   *Verbascum spp.: Scrophulariaceae*

The verbascums or mulleins are visited by honey bees at times, mainly for pollen. The majority, wild and cultivated, are in flower late in the summer and so furnish pollen when many other sources are over. When nectar is secreted it appears to be done so sparingly and on the inner sides of the petals (11). The so-called great mullein (*V. thapsus*) which grows 4 to 5 feet high and has light yellow flowers, is often to be seen on waste ground, especially on chalk or light soils, and is not uncommon as a wild plant. It sometimes attracts bees well for pollen.

Verbena   *Verbena spp.: Verbenaceae*

There are over 100 species of *Verbena*, annuals or perennials, and mostly natives of the New World. In many the flower-tubes are too long for the honey bee, but in others this is not so and as they often produce nectar freely they are good bee plants. This is the case with some of the blue-flowered, hardy perennial verbenas that have been introduced to cultivation in Britain, such as *V. hastata* and *V. stricta*, the latter being freely worked for nectar and pollen at Kew.

The only wild verbena in the British Isles is the vervain (*V. officinalis*), of great repute in bygone days for medicinal purposes. It is very common in England, but less so in Scotland and Ireland, and is often to be seen in quantity around rubbish heaps, along roadsides, the outskirts of villages, etc. Its pale lilac flowers appear in slender spikes in July and August and bees are frequent visitors for nectar.

Veronica   *Veronica spp.: Scrophulariaceae*

Bees visit the flowers of both the wild and the garden or cultivated veronicas. Among the latter *V. longifolia* (sometimes sold as *V. spicata*), which is the commonest garden sort, is an excellent bee plant. It is of sturdy and erect growth, with long flower spikes, and is in bloom a long time, hence its general popularity. Its flowers never fail to attract the honey bee in large numbers. This applies to the various varieties with blue, white, rose or purple flowers. The nectar, which may be produced copiously, is secreted by the fleshy disc at the base of the ovary and is protected by hairs in the throat of the flower tube. No bee garden should be without this excellent bee plant, which will thrive in any ordinary garden soil in a sunny position.

Vetch   *Vicia sativa: Leguminosae*

Vetches or tares are an important forage crop throughout the country and are extensively grown, being either spring or autumn sown. Unfortunately they do not rank as good or first-class bee plants, although bees visit the flowers on occasions and seem to get a certain amount of nectar. There are even records of surplus honey being obtained from them. This is said to resemble that of clover but to have a stronger flavour (23). It would seem that the plant may be useful for nectar in some seasons, but not in others and that some localities are more favourable than others.

Like its close relative the field bean (*V. faba*) the vetch has well-developed extra-floral nectaries. These become functional about a fortnight before the flowers open. As the flower is rather long for the honey bee it is not unusual for the nectar to be reached from the back or side (*American Bee Journal*, 1940, 213).

Besides the common or field vetch there are many wild vetches whose flowers may receive visits from the honey bee, such as the tufted vetch (*V. cracca*), bush vetch (*V. sepium*), horse-shoe vetch (*Hippocrepis comosa*) and the Siberian or hairy vetch (*V. hirsuta*, formerly *V. villosa*).

Vetchling   *Lathyrus spp.: Leguminosae*

The vetchlings are a rather similar group of plants to the vetches, scrambling or climbing by means of tendrils, and with winged or flattened stems like the sweet pea which belongs to the same genus. There are fewer leaflets per leaf than in the vetches. Although of prolific growth they are not so useful to the agriculturist as they are less radily eaten by stock and in some instances are suspected of poisonous properties.

The so-called Wagner pea (*L. sylvestris wagneri*) of Austria is probably the most interesting of this group of plants from the beekeeper's point of view, for the small pea-like flowers are worked most industriously by honey bees for nectar and are produced in great profusion. This plant is simply a cultivated form of the narrow-leaved everlasting or flat pea (*L. sylvestris*), wild in many parts of Britain and particularly abundant along some of the Cornish cliffs. It was first brought to light or developed by W. Wagner of Württemburg from the Carpathian Mountains in Austria in the latter part of last century, and is claimed to be much more palatable to livestock than the ordinary wild form. It is a long-lived perennial with stems 3 to 6 feet long and produces bunches of small pink flowers, three to ten in each bunch. The plant is slow in becoming established and flowering may not take place until the second season. When well established in a field, which requires two to three years, it forms a dense mass of vegetation 3 to 4 feet high and covered with flowers. In a small plot (spring sown) established by the writer a few plants flowered sparingly in the first season.

The Wagner pea has recently been grown experimentally in England and has proved most attractive to bees and obviously a good nectar yielder. It is hoped, in the interests of beekeeping, that further trials may lead to its being adopted in agriculture either as a fodder or forage plant, especially for poor, light soils as it is drought resistant. It may be useful as a green manure on account of the mass of vegetation it produces and its long taproot, or possibly as an orchard cover in fruit growing districts. The Wagner pea has attracted attention in the United States as a bee plant (*American Bee Journal*, 1941, 540–1).

Violet   *Viola odorata: Violaceae*

In warmer climates violets are commonly visited by honey bees for nectar. Among the many varieties in cultivation the length of the nectar-containing spur is variable and some are better suited for the hive bee than others. The wild violets of the hedgerows appear so early in the year that they are seldom likely to be of any consequence as a source of nectar. Bees also visit the flowers of the common wild viola or pansy (*V. tricolor*) but it is probably of little account as a bee plant. However, its pollen has been found in honey (27).

Viper's Bugloss   *Echium vulgare: Boraginaceae*

The conspicuous purple flowers of this plant are well known for their attraction to bees. Without doubt the viper's bugloss is one of the

most stately of British wild plants and one of the most beautiful. Its
rough stem is speckled (like a viper) and grows from 1 to 3 feet in
height. The narrow leaves are also rough or bristly.

The plant sometimes occurs as a weed in and around fields but is
most frequently seen on chalky hills and sea cliffs. It has become a
bad weed in some parts of Canada. Unfortunately for the beekeeper
the plant is somewhat local in its distribution in Britain. Flowering
takes place mainly in June and July. The colour of the flowers is
somewhat variable and in addition to purple may be pale blue,
bluish pink or even white. The plant sometimes finds a place in the
flower garden, although other echiums are more generally cultivated.

Nectar is produced freely in the flowers of this plant and it has
frequently been recommended for sowing in waste places to improve
the bee pasturage of a district. It was one of the plants selected for
sowing on the cuttings and embankments of roads and railways in
Germany for this purpose (*Bee World*, 1935, 105).

A similar plant, the Jersey bugloss (*E. lycopsis*, formerly *E.
plantagineum*) a European species which occurs in Jersey and parts
of Cornwall is also reputed to be a good bee plant. It has become a
common weed in Australia, especially South Australia, where it may
cover acres of wheat land. Commonly known as blue weed or
Patterson's curse, it has a reputation there of being a useful source of
nectar and pollen.

Virginia Creeper   *Parthenocissus quinquefolia: Vit acea*

This climber is very commonly cultivated. It is to be seen covering
walls and the sides of houses everywhere and is valued for the
brilliant red colour it turns in the autumn just before the leaves fall.

Its flowers are in small bunches of three to eight and are not at all
conspicuous, especially when the plant grows on walls, for then they
are completely hidden from view by the overhanging leaves. Their
presence, however, is usually given away by the merry hum of bees
as they seek the flowers for nectar and for pollen.

The Virginia creeper is regarded as a useful nectar plant in its
native land (eastern North America), where it frequently climbs the
loftiest trees. In the mid lands of Natal, where the creeper is favoured
on houses for keeping them cool in the hot weather, the writer has
many times observed it when in flower swarming with bees.

Wallflower   *Cheiranthus spp.: Cruciferae*

Wallflowers are among the most useful of the early spring garden flowers for bees and are visited constantly for nectar and pollen when weather is suitable. There seems to be little if any difference between the many varieties in cultivation where attraction to bees is concerned. All are popular except of course the double-flowered varieties and these are not grown to nearly the same extent as the single forms.

The pollen is pale or greenish yellow in colour as a rule, the individual grain being rather small (about 18 microns) and rough in outline. The nectaries take the form of two swollen ridges at the base of the two short stamens.

The wallflower is not really a native of Britain, being of European origin, but is to be found naturalized or apparently wild in some parts of the country, especially on cliffs near the sea, old walls, and rocky places generally.

Walnut   *Juglans regia: Juglandaceae*

The walnut is of little consequence to bees. It produces no nectar, being a wind-pollinated tree. The pollen is light and powdery as is usual in plants of this kind—the type of pollen bees do not usually care about when other types of pollen are available. As the walnut produces its tassels of male flowers fairly late in the season, when there is an abundance of other pollen, it is not usual for bees to visit the blossoms to any extent in Britain. In other countries, however, bees have been known to work the flowers zealously for pollen (23).

Watercress   *Nasturtium officinale: Cruciferae*

The flowers of watercress, like those of most cruciferous plants, are a source of nectar to the hive bee and are usually available from June till August. The plant is common on the margins of streams and ponds. It is also cultivated on a large scale for market in special watercress beds, sometimes several acres in extent. However, normally it is harvested before it reaches the flowering stage. The water-rocket (*N. amphibia*) is a somewhat similar plant with yellow flowers and is also visited for nectar.

Weigela (formerly **Diervilla**)   *Weigela spp. (formerly Diervilla):*
                                                       Caprifoliaceae

These deciduous shrubs, which are allied to the shrubby honeysuckles, secrete nectar very freely. They are among the most beautiful
of summer-flowering shrubs with white, cream, pink, or red flowers
of various shades. Bumble bees may be seen working the flowers with
zest in June and frequently bite holes at the base of the flower which
honey bees make use of to obtain the nectar.

Whitlow Grass   *Erophila verna: Cruciferae*

The small white flowers of this wild British plant, often to be seen on
old walls and in dry, rocky places, are sometimes visited by bees for
nectar and pollen.

Willow   *Salix spp.: Salicaceae*

The willows or sallows are useful plants to the beekeeper in the early
part of the season. There are some twenty wild species in the British
Isles as well as numerous hybrid forms. Other introduced kinds, like
the weeping willow (*S. babylonica*), are also often cultivated. The
willows vary in size from large trees, like the crack willow (*S. fragilis*)
and white willow (*S. alba*) to small shrubs, sometimes not more than
a few inches high and of a creeping nature. Some, like the common
sallow (*S. cinerea*) and goat willow (*S. caprea*)—commonly called
palm and much sought for decoration when in the silvery bud stage—
and the dwarf or creeping willow (*S. repens*) are widely distributed
throughout the country and very prevalent in some districts. Their
presence is always appreciated by local beekeepers as a source of
early pollen and nectar.

Willows are in flower from March to May according to species and
district. The male and female catkins are borne on different plants
in most cases. These vary much in size and shape and are generally
silky, the minute flowers constituting them being devoid of petals
and consisting only of a scale bearing the stamens or pistil in the case
of the female catkin. Small yellow glands or nectaries may also be
present. Many willows are a source of nectar as well as pollen to the
honey bee when weather conditions are favourable. Catkins generally
appear on the naked shoots of the previous summer and before the
leaves arrive. Both male and female flowers secrete nectar under
favourable conditions.

The willows already mentioned are known to be visited by honey bees for nectar and pollen (11) but there are many more. Included among them are the round-eared willow (*S. aurita*) found in woods, almond-leaved willow (*S. triandra*) an osier cultivated for basket-making, and the violet willow (*S. daphnoides*). The last mentioned is a European willow, although naturalized in Yorkshire, and has been recommended on the Continent for cultivation by beekeepers, for it flowers earlier than most willows and has large catkins (*Bee World*, 1936, 138). Furthermore, it is highly ornamental with its waxy, purple shoots.

Willows in close proximity to an apiary are always desirable. Most willows are easily propagated by cuttings placed in the open ground any time between November and early March. In the case of the tree sorts like the crack willow and white willow, quite large sets, 6 to 10 feet long and 1 to 1½ inches in diameter usually root readily and are commonly used. If difficulty is experienced in obtaining rooting, as may happen in dry localities, the following is a more certain method of obtaining rooted plants. Cuttings about a foot long and the thickness of a pencil are taken in the winter and kept in a cool place or heeled in. In March these are placed with their ends in water (half immersed) and brought indoors or placed in a greenhouse when the warmth and moisture soon causes profuse root development. The rooted cuttings may then be planted out in April or May and kept watered for a time in the event of dry weather.

The male forms of the goat willow (*S. caprea*) with their large handsome catkins are probably as good as any for cultivation by the beekeeper. Some forms flower earlier than others and could be used to lengthen the flowering period. Honey may be obtained from willow in seasons when a fine spell of weather accompanies flowering, as in 1945.

Winter Aconite   *Eranthis hyemalis: Ranunculaceae*

This tuberous-rooted perennial, with its attractive yellow flowers produced so early in the year (January to March), is not a native plant but originated from western Europe. It is naturalized in woods here and there and is much favoured for shrubberies as it does not exceed 6 to 8 inches in height and flowers before anything else. In mild winters the flowers are out in January and are always well ahead of the crocus. With suitable weather bees work them industriously, not only for pollen but for nectar as well.

In this flower the nectar is produced and stored in the same way as in the Christmas rose (*Helleborus niger*), in vase-shaped containers arranged on the flower where the petals are usually to be found. They are in fact modified petals.

Like the crocus the winter aconite is well worth cultivating near hives as an early source of pollen. Once established it requires little attention and grows well under shrubs or trees where few other plants succeed. It is usually in flower from four to six weeks.

Winter Cress   *Barbarea vulgaris: Cruciferae*

This plant occurs both wild and cultivated. Its yellow mustard-like flowers appear in dense terminal clusters from May to June. Bees visit them for nectar.

Wisteria   *Wisteria sinensis: Leguminosae*

The purple masses of bloom of this magnificent creeper will attract the honey bee in some seasons but not in others. They appear towards the end of May as a rule and there is sometimes a second but much smaller crop of flowers in August. In bright or warm weather bees visit the flowers for nectar at Kew but when conditions are cool they seem to offer little attraction. In other warmer climates the flowers of wisteria are known to be worked for nectar and pollen (23).

Woad   *Isatis tinctoria: Cruciferae*

The woad plant of the Ancient Britons is to be found wild in old stone pits and chalk quarries in some parts of the country. Its loose clusters of yellow flowers attract bees for nectar but the plant is nowhere sufficiently abundant to be of any consequence to the beekeeper. It was cultivated as a dye plant at one time but its cultivation in England was finally given up some years ago.

Wood Sage   *Teucrium scorodonia: Labiatae*

Wood sage or wood germander as it is sometimes called grows in masses in woods, in heathy places and in other situations. It is 1 to 2 feet high and has wrinkled leaves very like those of ordinary sage. Its yellowish-green flowers appear in July and August. They often have a faint tinge of purple and grow in terminal one-sided clusters. The flower-tube is some 9 to 10 mm. in length and as nectar is secreted freely and may reach a level of 4 to 5 mm. up the tube the honey bee is able to obtain it. Bees often work the flowers very freely in fact, and surplus has been obtained from the plants in parts of the country where the plant is common (9).

Some of the other species of *Teucrium* attract bees, including the water germander (*T. scordium*), which, however, is not a common plant.

Wormwood   *Artemisia absinthium: Compositae*

This strong-smelling plant is repulsive rather than attractive to bees but is nevertheless of interest to the beekeeper on this account. Some have recommended rubbing the hands with the plant when manipulating to avoid stings. It is also claimed that it will keep away wax moth from stored combs for all insects dislike it, and that in the case of robbing a few crushed twigs placed in front of the hive are likely to be effective. When a swarm has settled in an inconvenient place, as between the branches of trees or bushes it may be induced to move by stroking it with bruised branches of wormwood (*Bee World*, 1931, 79).

Yellow-wood   *Cladastris lutea* (formerly *C. tinctoria*):
                                                          Leguminosae

This interesting tree from the south-eastern United States is not often cultivated in Britain and furthermore does not flower regularly. When it does produce flowers they appear in June and are in large bunches 1 foot or more long. They are white and slightly fragrant. Bees visit them at Kew in some seasons. In its native land the tree is the source of some surplus honey (19).

Yew   *Taxus baccata: Taxaceae*

The yew is common both wild and cultivated, some of the largest trees being those in churchyards and these are often of great age. It is most prevalent in the chalk districts of south-east England. On the western Sussex Downs it forms extensive woods. Yew flowers early, generally in March, and produces pollen in abundance from it male flowers. This is light and powdery and will often float away in clouds on the branches being tapped. Bees work the flowers well for pollen in some districts. Possibly, as is the case with other dry or wind-borne pollens, bees only take it when no other source is available. The yew does not yield nectar.

Bibliography

(Numbers in square brackets following certain references are of the relevant pages in the text.)

(1) Alefeld, F., *Die Bienenflora Deutschlands und der Schweiz* (Neuwied, 1863). [131, 135]

(2) Armbruster, L. and Oenike, G., *Die Pollenformen als Mittel zur Honigherkunftsbestimmung* (Neumunster, 1929).

(3) Browne, C. A., *Chemical Analysis and Composition of American Honeys* (U.S. Dept. Agric. Bull. No. 110, 1908). [184]

(4) Delaigues, Abbé, *Les Abeilles et les Fleurs: Plantes Mellifères* (Paris, 1926). [104, 131, 163, 179, 185, 207]

(5) Edgworth, M. P., *Pollen* (London, 1887).

(6) Erdtman, G., *An Introduction to Pollen Analysis* (Altham, Mass., U.S.A., 1943). [30]

(7) Goodacre, W. A., *Honey and Pollen Flora of New South Wales* (Sydney, 1938).

(8) Hayes, G., *Nectar Producing Plants and their Pollen* (London, 1925). [32, 93, 117, 148]

(9) Herrod-Hempsall, W., *Beekeeping New and Old*, Vols. I (1930) and II (1937) (London). [56, 82, 83, 85, 86, 91, 111, 205, 218]

(10) Howes, F. N., *Bee Plants in Surrey* (Kew Guild Journal, 1943).

(11) Knuth, P., *Handbook of Flower Pollination* (Oxford, 1906, 1908). [101, 113, 127, 131, 133, 137, 144, 145, 167, 168, 170, 183, 184, 210, 216]

(12) Lovell, J. H., *The Flower and the Bee* (London, 1919).

(13) Mace, H., *Bee Farming in Britain* (Beekeeping Annual Office, Harlow, Essex, 1936). [56]

(14) Manley, R. O. B., *Honey Production in the British Isles* (London, 1936). [17, 19, 57, 84, 90, 125, 166, 169, 187, 188, 195]

(15) Melzer, H., *Bienen-Nährpflanzen* (1894).

(16) Müller, H., *The Fertilization of Flowers* (Leipzig, 1883). [183, 198]

(17) Pammel, L. H. and King, C. M., *Honey Plants of Iowa* (Des Moines, Iowa, U.S.A., 1931).

(18) Parker, R. L., *The Collection and Utilization of Pollen by the Honey Bee* (Cornell Univ. Agric. Exp. Stn., Memoir No. 98, 1926). [28]

(19) Pellett, F. C., *American Honey Plants* (American Bee Journal, Hamilton, Illinois, U.S.A., 3rd edition, 1930). [109, 120, 126, 132, 136, 139, 140, 162, 167, 169, 175, 184, 192, 197, 201, 203, 210, 218]

(20) Root, A. I. and E. R., *The A.B.C. and X.Y.Z. of Bee Culture* (Medina, Ohio, U.S.A., 1919 edition). [24, 30, 52, 60, 112, 121, 128, 147, 155, 181, 198, 202, 208]

(21) Scullen, H. A. and Vansell, G. H., *Nectar and Pollen Plants of Oregon* (Agric. Exp. Stn., Corvallis, Oregon, 1943).

(22) Tinsley, J., *Experimental Work in the Production of Heather Honey* (West of Scotland Agric. College, Bull. No. 144, Glasgow, 1943). [17, 71]

(23) Vansell, G. H., *Nectar and Pollen Plants of California* (Berkeley Agric. Exp. Stn., Univ. of Calif., Bull. No. 517, 1931). [83, 98, 104, 119, 123, 124, 135, 137, 139, 145, 165, 168, 173, 175, 197, 201, 211, 214, 217]

(24) Von Tafel, Dr., *Der Bienengarten* (1942).

(25) Wedmore, E. B., *A Manual of Beekeeping* (London, 1932). [157]

(26) Wodehouse, R., *Pollen Grains* (New York, 1935).

(27) Yate-Allen, Rev. M., *European Bee Plants and their Pollen* (Alexandria, Egypt, 1937). [66, 128, 130, 140, 141, 142, 146, 147, 151, 153, 161, 166, 174, 180, 194, 196, 200, 206, 207, 212]

(28) Zander, E., *Pollengestaltung und Herkunftsbestimmung bei Blütenhonig* (Berlin, 1935).

(29) *Report on the Marketing of Honey and Beeswax in England and Wales* (H.M. Stationery Office, London, 1931). [20, 62]

(30) Pryce-Jones, J., *Some Problems Associated with Nectar, Pollen and Honey* (Linnean Society, London, 1944. Extracted from Proceedings Session 155, pp. 129–74). [20, 72, 74, 201, 208]

Supplementary Bibliography

prepared by Dr Eva Crane

(Numbers in square brackets following certain references are of the relevant pages in the text.)

The references provided here will give readers of the 1979 reprinted edition access to information published since 1945.

Several interesting publications describe how honey bees find and forage on plants, for instance 37, 38. One hazard not known in 1945 was the possible contamination of nectar by systemic insecticides, see 45, 46.

Most publications listed below were in print in January 1979. They can be purchased through the International Bee Research Association, Hill House, Gerrards Cross, Buckinghamshire SL9 0NR (except 41, 45–7, 51–5, 59, 60, 64). These, and Dr Howes's items 1–30 (except 1, 10, 15, 16, 21, 24, 26), are all in the IBRA Library.

(31) Bee World, *Pollen and its harvesting* (International Bee Research Association Reprint M86, 1976). [26]

(32) Crane, E., *Bees in the pollination of seed crops* (Journal of the Royal Agricultural Society of England, 133, 119–35, 1972). [32]

(33) Crane, E., *Honey: a comprehensive survey* (London, Heinemann and Bee Research Association, 1975; reprinted with corrections, 1976). [13, 19, 31].

(34) Dadant & Sons, *The hive and the honey bee* (Hamilton, Ill., Dadant & Sons, 1975) especially Chapter VIII by R. W. Shuel, *The production of nectar.* [13]

(35) Deans, A. S. C., *Survey of British honey sources* (London, Bee Research Association, 1957). [19, 55]

(36) Free, J. B., *Insect pollination of crops* (London, Academic Press, 1970). [32]

(37) Frisch, K. von, *The dancing bees* (London, Methuen, 2nd edition, 1966).

(38) Frisch, K. von, *The dance language and orientation of bees* (Cambridge, Mass., Harvard University Press, 1967).

(39) Ghisalberti, E. L., *Propolis: a review* (International Bee Research Association Reprint M99, 1979). [51]

(40) Gleim, K.-H., *Nahrungsquellen des Bienenvolkes* (St Augustin, German Federal Republic, Delta Verlag, 1977). [18, 55]

(41) Hillyard, T. N. and Markham, J., *A survey of beekeeping in Ireland* (Dublin, An Foras Talúntais, 1968). [55]

(42) Hodges, D., *A calendar of bee plants* (Bee World, 39, 63–70, 1958; also IBRA Report M94, 1978). [96]

(43) Hodges, D., *The pollen loads of the honeybee* (London, Bee Research Association, 1952; facsimile reprint with addenda, 1974). [27]

(44) International Commission for Bee Botany of IUBS, *Methods of melissopalynology* (International Bee Research Association Reprint M95, revised edition 1978). [27]

(45) Jaycox, E. R., *Effect on honey bees of nectar from systemic insecticide-treated plants* (Journal of Economic Entomology, 57, 31–5, 1964).

(46) King, C. C., *Effects of herbicides on honeybees* (Gleanings in Bee Culture, 92, 230–3, 251, 1964).

(47) Kloft, W., Maurizio, A. and Kaeser, W. *Das Waldhonigbuch* (Munich, Ehrenwirth Verlag, 1965). [50]

(48) Louveaux, J., Maurizio, A. and Kirkwood, A. M. C., *Heather: bees—honey—beekeeping* (International Bee Research Association Reprint M71, 1973). [70]

(49) MacGregor, J. L., *Poisoning of bees by rhododendron nectar* (Scottish Beekeeper, 36, 52–4, 1960). [23]

(50) McGregor, S. E., *Insect pollination of cultivated crop plants* (Agricultural Handbook, U.S. Department of Agriculture No. 496, 1976). [32]

(51) Maurizio, A., *Blüten, Nektar, Pollen, Honig* (Deutscher Bienenwirtschaft, 10 and 11, 1959 and 1960). [18, 31]

(52) Maurizio, A. and Grafl, I., *Trachtpflanzenbuch* (Munich, Ehrenwirth Verlag, 1969). [18, 31]

(53) Meeuse, B. J. D., *The story of pollination* (New York, Ronald Press, 1961). [32]

(54) Melville, R., *The limes as amenity trees and bee pasturage* (Kew Bulletin No. 2, 147–51, 1949). [62]

(55) Meyer, W., *Propolis bees and their activities* (Bee World, 37, 25–36, 1956). [51]

(56) Mountain, M. F., *Trees and shrubs valuable to bees* (London, Bee Research Association, revised edition, 1975). [39, 41, 45, 47]

(57) Mountain, M. F. and others, *Garden plants valuable to bees* (London, International Bee Research Association, 1979). [41, 42, 45, 47]

(58) Pellett, F. C., *American honey plants* (Hamilton, Ill., Dadant & Sons, 1945, reprinted 1976). [38, 41, 45, 47]

(59) Percival, M., *Types of nectar in Angiosperms* (New Phytologist 60, 235–81, 1961). [14]

(60) Percival, M., *Floral biology* (Oxford, Pergamon Press, 1965). [14, 31]

(61) Procter, M. and Yeo, P., *The pollination of flowers* (London, Collins, 1973). [32]

(62) Stanley, R. G. and Linskens, H. F., *Pollen: biology—biochemistry —management* (Berlin, Springer-Verlag, 1974). [26]

(63) Wedmore, E. B., *A manual of bee-keeping for English-speaking bee-keepers* (facsimile reprint of 1946 edition, Warminster, U.K., Bee Books New & Old, 1975). [96]

(64) White, J. W., *Toxic honeys* (Pages 495–507 from *Toxicants occurring naturally in foods*, National Research Council, 2nd edition; Washington, National Academy of Sciences, 1973). [23]

General Index

NOTE—The first page reference after each entry in the index indicates the page on which the matter is dealt with in greatest detail

Tamarisk, 202, 44, 49
Tannehonig, 52
Tansy, 202
Tares, 211
Teasel, 203, 18, 38, 47
Tea tree, 203
Thalictrum, 191
Thistle, 203
Thrift, 204, 44
Thyme, 205, 44, 45; wild, 205
Toad-flax, 206
Tobacco, 206; mountain, 103
Tomato, 206
Traveller's joy, 125
Tree: honey, 19, 80; mallow, 163
Tree-of-heaven, 207
Trefoil, 61, 166
Trillium, 207
Tulip, 207, 44
Tulip-tree, 208, 13, 15
Tupelo, 208, 15
Turnip, 209, 113

Valerian, 209
Vegetable marrow, 209, 148
Vegetables, 44
Verbascum, 210
Verbena, 210
Veronica, 210, 44
Vervain, 210
Vetch, 211
Vetchling, 211
Viola, 212
Violet, 212
Viper's bugloss, 212, 38

Virginia creeper, 213
Vitamins, 20

Wagner pea, 211
Wallflower, 214
Walnut, 214
Wasps, 32
Water: avens, 145; mint, 174; pepper, 184
Watercress, 214
Wax, 29
Weigela, 215
Whinberry, 108
Whitethorn, 84
Whitlow grass, 215
Whortleberry, 108
Willow, 36, 47, 50, 215
Willow-herb, 88, 15, 39, 47; hairy, 90
Windbreaks, 47
Winter aconite, 216, 17, 44, 46, 153
Winter cress, 217
Winter heliotrope, 117
Wisteria, 217
Woad, 217
Wolf-berry, 198
Woodbine, 156
Wood: lily, 44; rush, 196
Wood sage, 217
Wormwood, 217
Wound-wort, 199

Yellow rattle, 151
Yellow-wood, 217
Yew, 217, 49

Index of Genera